Routledge Revivals

The English Medieval Landscape

First published in 1982, *The English Medieval Landscape* was written to recreate and analyse the development of the major elements of the medieval landscape.

Illustrated with maps and photographs, the book explores the nature of the English landscape between 1066 and 1485, from farms and chases to castles, monastic settlements, villages, roads, and more.

The English Medieval Landscape will appeal to those with an interest in medieval history and British social history.

The English Medieval Landscape

Edited by Leonard Cantor

Routledge
Taylor & Francis Group

First published in 1982
by Croom Helm Ltd.

This edition first published in 2021 by Routledge
2 Park Square, Milton Park, Abingdon, Oxon, OX14 4RN
and by Routledge
605 Third Avenue, New York, NY 10017

Routledge is an imprint of the Taylor & Francis Group, an informa business

© 1982 L. M. Cantor

All rights reserved. No part of this book may be reprinted or reproduced or utilised in any form or by any electronic, mechanical, or other means, now known or hereafter invented, including photocopying and recording, or in any information storage or retrieval system, without permission in writing from the publishers.

Publisher's Note
The publisher has gone to great lengths to ensure the quality of this reprint but points out that some imperfections in the original copies may be apparent.

Disclaimer
The publisher has made every effort to trace copyright holders and welcomes correspondence from those they have been unable to contact.

A Library of Congress record exists under LCCN: 82128514

ISBN 13: 978-0-367-74826-5 (hbk)
ISBN 13: 978-1-003-15938-4 (ebk)
ISBN 13: 978-0-367-74754-1 (pbk)

Book DOI: 10.4324/9781003159384

The English Medieval Landscape

EDITED BY LEONARD CANTOR

CROOM HELM
London & Canberra

© 1982 L.M. Cantor
Croom Helm Ltd, 2–10 St John's Road, London SW11

British Library Cataloguing in Publication Data

The English medieval landscape. – (Croom Helm
    historical geography series)
    1. England – Rural conditions – History
    2. England – Medieval period, 1066–1485
    I. Cantor, Leonard
    942'.00973'4   DA667
    ISBN 0-7099-0707-9

Typeset by Leaper & Gard Ltd, Bristol

Printed and bound in Great Britain
by Billing and Sons Limited
Guildford, London, Oxford, Worcester

CONTENTS

FIGURES AND TABLES

Figures

Tables

PLATES

PREFACE

This book is an attempt by a group of historical geographers to recreate and to analyse the development of the major elements in the English landscape during the Middle Ages, a period of over four hundred years from 1066 to 1485. There are, of course, already in existence a number of authoritative books which deal, wholly or in part, with the historical geography of medieval England. Amongst those which have appeared in the past decade are H.C. Darby, *A New Historical Geography of England Before 1600* (Cambridge, 1976); R.A. Dodgshon and R.A. Butlin, *An Historical Geography of England and Wales* (Academic Press, 1978); E. Miller and J. Hatcher, *Medieval England – Rural Society and Economic Change, 1086-1348* (Longman, 1978); and M.M. Postan, *The Medieval Economy and Society: An Economic History of Britain in the Middle Ages* (Wiedenfeld and Nicolson, 1972). As the titles of these books imply, they are primarily concerned with giving a synoptic view of the geography or economic history of the country, usually by reference to specific chronological dates or periods such as Domesday England, the Early Middle Ages and the Later Middle Ages. However, this book adopts a different approach: it is our belief that an equally valid and valuable way of analysing the development of the English landscape over the long period of the Middle Ages is by identifying and describing in detail the major elements which constituted it. Fortunately, in recent years much detailed research has been undertaken by historical geographers and others, producing a great deal of information and evidence that now make it a practical proposition to draw up a reasonably accurate, if partial, picture of the development of the English landscape during the medieval period.

The authors attempt both to give a synoptic view of the country as a whole and also to describe in detail specific examples and case studies of the various elements, to illustrate the general theme and to identify regional variations. They also describe the wide range of techniques now available to students, whether professional or amateur, wishing to recreate elements of past landscapes. It was intended to include a chapter on the medieval industrial landscape; unfortunately, however, circumstances beyond the editor's control and considerations of length made this impossible.

Finally, when making reference to English counties, we have taken as their boundaries those which existed before the changes in April 1974; these changes were largely the product of administrative convenience with little regard for history, and the vast majority of documentary and other material upon which we have drawn refers to the traditional county boundaries.

ACKNOWLEDGEMENTS

I wish to thank Professor Robin Butlin, Department of Geography, Loughborough University, for his help in bringing this book to fruition. I am also grateful to F.A. Aberg of the Moated Sites Research Group for generously supplying information and the map on moated home-steads, and especially to Janice Oselton for the enormous amount of typing and retyping that was necessary before the typescript achieved its final form. We are grateful also to Mrs Anne Tarver, cartographer in the Loughborough University Department of Geography, for drawing the maps that illustrate Chapters 3 and 5, to Steve Chilton, carto-grapher at Middlesex Polytechnic, for the maps in Chapter 6, and to Gustav Dobrzynski, cartographer in the Department of Geography, University of Salford, for the maps in Chapter 7.

Acknowledgement is given to Aerofilms for Plate 1; to the Director, Cambridge University Collection of Aerial Photographs for Plates 4, 5, 6, 7 and 8; to David Popham and the owners of Kingston Lacy House, Dorset for Plate 2; and to Joe Tarrant for Plates 3a and b. Above all I am grateful to my wife for so tolerantly and encouragingly enduring the protracted labours associated with the writing of this book.

L.M. Cantor
Loughborough

1 INTRODUCTION: THE ENGLISH MEDIEVAL LANDSCAPE

Leonard Cantor

By the time the Normans arrived in England in 1066, the major characteristics of the English landscape had already been determined by their
Anglo-Saxon and Scandinavian predecessors. The settlement pattern of
towns and villages, the clearing of the woodlands, and the development
of agriculture had assumed a pattern which the Normans were to
modify but not substantially alter. In Darby's words, 'the Norman
Conquest was the transposition of an aristocracy and not a folk movement of new settlers on the land'.[1] In 1066, the Normans instituted a
survey of their conquered country which resulted in the Domesday
Book. Although incomplete and imprecise, for example it omits to deal
with the four northern counties, it is nevertheless an immensely important and interesting document which enables us to obtain a reasonably
accurate view of the nature of the English countryside in the late
eleventh century. In this respect, in common with all scholars of
Domesday England, we are enormously indebted to the work of H.C.
Darby and his colleagues.[2] The picture that emerges from the Domesday survey is of a country already possessing a well-developed territorial
organisation at the time of the Conquest, divided into shires, which in
turn were subdivided into hundreds or wapentakes consisting of vills
or villages, of which about 13,000 are individually named in Domesday.
The land was already settled and tilled to a remarkable extent; indeed,
it has been conservatively estimated that the arable acreage in 1086 was
over 80 per cent of the total area still under the plough as recently as
1914.[3] The areas of the country with most arable land were largely
those where soils were favourable including the Sussex plain; eastern
East Anglia; parts of the Midlands such as north Oxfordshire, the Vale
of Evesham, south-east Hertfordshire and south Warwickshire; and the
lower Exe Basin in Devon.[4] Areas with relatively little cultivated land
included the Weald, the New Forest, the Dorset and Surrey heathland,
the Fenlands and the Breckland of East Anglia, together with the
northern counties which had been deliberately devastated by the
Conqueror.

Rackham[5] estimates that by 1086 about 35 per cent of England
was devoted to arable land, at least 25 per cent to pasture, 15 per cent

17

to woodland and the remaining 25 per cent to settlements, moorland, fens, heaths and land devastated by war. Certainly, grassland was scattered throughout the country and virtually every village possessed some meadow. The woodland cover was extensive and, in addition to common woodland which was present in almost every manor, royal hunting forests had come into being in the twenty years since the Conquest. Although the latter were areas legally defined by the newly introduced forest law and not necessarily 'forests' in the geographical sense, they were often well wooded. In addition, at least 36 small enclosed hunting grounds in the form of parks had been created since 1066. Marshland was extensive, with large tracts in Romney Marsh, the Isle of Axeholme in Lincolnshire, East Anglia, Essex and Somerset and, by inference, along other stretches of coast and up the river valleys. Wasteland was also an important element in the early Medieval land-scape, though the Domesday term 'waste' does not connote the natural waste of mountain, heath and marsh, but land that had gone out of cultivation mainly as a result of deliberate devastation. Most of this devastation was wrought by the Conqueror's armies, especially in the north and, to a somewhat lesser extent, in the west of the country, though by 1086 the countryside was on its way to recovery.[6]

There was a well-developed settlement pattern and villages were very numerous. Towns were clearly in existence, but the evidence in Domes-day is so incomplete and unsystemmatic that it is very difficult to form a clear idea of their size and disposition. London must have been the largest town with a population perhaps in excess of 10,000 and next in size, though probably with little more than 5,000 inhabitants each, came Winchester, York, Lincoln, Norwich and, possibly, Thetford in Norfolk.[7] The total population of England at the time of Domesday has been variously estimated as between 1.1 million and over 2 million, mostly concentrated in the east and south-east of the country, especi-ally East Anglia and Kent.[8] Industry was relatively unimportant and consisted very largely of some iron-working, lead-working in the Derby-shire Peak District, stone-quarrying and salt-making.

From the time of the Conquest until the beginning of the fourteenth century — a period of about 250 years commonly termed the Early Middle Ages[9] — England generally experienced economic expansion and prosperity. This is well illustrated by the growth in population which occurred during this period, from about 1½ million in 1086 to perhaps 4 million in 1300.[10] This substantial rise in population inevit-ably made a major imprint on the English landscape and, in the country-side, the cultivated area was steadily extended, principally by the reclamation of hitherto marginal areas. Postan identifies three main categories of marginal land which were brought under cultivation at this time. The first comprised the marshes and fens, including Romney Marsh, Sedgemoor in Somerset and especially the Fenlands, which, once drained, provided very fertile land. The second was the royal forest, generally in areas of relatively poor land, which from about 1185 was increasingly 'disafforested', that is removed from the jurisdiction of the forest law. The third consisted of the peripheral regions of the country, in the far north, the south-west and along the Welsh border.[11] In addition, within the more densely cultivated areas of the south and the Midlands, the poorest tracts of heavy clay and thin chalk downland began to be farmed.

Such diagnostic features of the medieval landscapes as buildings and parks also greatly increased in number during this period. The first wave of castle-building followed the Conquest when the king and the Norman barons erected them as strongholds to hold down and garrison the lands they had seized. From about 1200 onwards, the castle, the fortified house and the moated homestead were built in substantial numbers, primarily as residences for the increasingly wealthy aristocratic and middle classes. Religious houses also proliferated as the monastic orders established themselves in the country, particularly during the so-called 'renaissance' of the twelfth century. The disafforestation of woodland and the prosperity of local landowners resulted in the creation throughout the country of many hundreds of hunting parks. This was also the great period of urban expansion, especially during the century 1150–1250, when towns flourished and multiplied and many boroughs were founded. Finally, industry also flourished: pottery manufacture and leather-tanning were widespread, metal-working prospered, as did mineral-working, glass-making and salt production.[12]

By contrast, the beginning of the fourteenth century ushered in a period of crisis, marked primarily by an agricultural slump which, with only a slight halt around 1410, continued until the end of the Middle Ages.[13] The reasons for the onset of this decline, which was sharply accelerated by the Black Death of 1348 and later plagues, are complex and the subject of much analysis and discussion.[14] However, a number of interlocking and deleterious changes were taking place from about 1300 onwards, including over-population, soil exhaustion, periods of persistently hostile weather, plagues, and royal taxation and debasement of the coinage.

As we have seen, by the fourteenth century the population of

England had probably expanded more than twofold to approaching 4 million, a growth which had outstripped the means of subsistence, for the labouring classes who constituted the great majority of the population. Land hunger during the thirteenth century had pushed out the bounds of cultivation into marginal lands with relatively poor soils, and by the end of the century soil exhaustion and poor yields were becoming more and more common. As a result, large-scale reclamation first slowed down, then came to a halt and, according to Postan, after 1350 and during the fifteenth century, the area under the plough began to contract.[15] This combination of declining yields and a slowing-down of the expansion of the cultivated land resulted in less agricultural produce being available for consumption by the peasant, bringing with it more malnutrition and a greater susceptibility to famine and plague. The decline in agricultural production was also occasioned, in part, by adverse climatic changes. The first half of the fourteenth century witnessed a period of unusual climatic instability with colder, more rainy weather than had previously been experienced. The climate was at its worst during the so-called 'agrarian crisis' of 1315–1322, which included the harvest failures of 1315, 1316 and 1321 and the great sheep murrain of 1319–21. Among the effects of the crisis were the abandonment by some landowners of demesne cultivation and the sale of marginal land by peasants.[16]

A major contributory factor to the fourteenth-century decline was the calamity of the Black Death in 1348. Though the first and greatest of the plagues, it was by no means the last, and other major pestilences affected England in the 1360s and 1370s and indeed later. While it is impossible to gauge precisely the rate of mortality brought about by the Black Death, a likely figure is thought to lie between 30 and 45 per cent of the total population,[17] and its chief effect was to delay the full economic recovery of the country until the end of the Middle Ages.[18] Indeed, it was not until the resurgence and growth of the late fifteenth and early sixteenth centuries that the total population of England reached its pre-Black Death figure.[19] Finally, the combined effects of increasing royal taxation and manipulation of the currency, made necessary by the desperate need of the Crown to raise money to wage war, brought distress to many local communities which was probably quite as great as famine and pestilence and fell most heavily on the peasantry.[20]

As far as the English countryside was concerned, the effect of these misfortunes was to bring about four major changes, which began to manifest themselves fully from the second half of the fourteenth century onwards: a decline in demesne farming, an increase in peasant farming, a change in land use from arable to pastoral, and the abandonment of settlement.

The decline in population brought about by the plagues and other causes inevitably resulted in a shortage of labour and, as a consequence, the second half of the fourteenth century was marked by a sharp rise in agricultural wages which became permanent and only levelled out after about 1370.[21] This led to increased costs of production and declining profits, especially for large landowners, who in many cases reacted by reducing the areas of the cultivated demesne lands and leasing them out to the peasants. This process, which preceded the Black Death by some decades, gradually increased until, by about 1450, almost all direct demesne farming had disappeared.[22] This development, whereby landlords increasingly became rentiers, was paralleled by the emergence of a class of richer peasant farmers who took up the leases and often found themselves acquiring fertile demesne land situated on good soil which had been well looked after. As the costs of labour increased, without a corresponding rise in market prices, landlords also resorted to turning from arable to pastoral farming. The rearing of cattle and sheep required less labour than arable farming, and wool production was fairly profitable.[23] This brought about a marked change in the appearance of the countryside, with a reduction in the area given over to the common fields and an increase in enclosure by hedges. The enclosure movement which long pre-dated the fourteenth-century recession was certainly accelerated by it and in Staffordshire, for example, the process increased after 1350 so that, by 1500, the amount of enclosed land probably equalled that of the common fields.[24] In Dorset, the common fields that disappeared seem to have been, in the main, those belonging to settlements which had become deserted or much reduced in size; in the large villages with no marked reduction in population the common fields remained. In Dorset, too, the appearance of the high chalk downland changed during this period as the arable land, established in the twelfth and thirteenth centuries, acquired the grassland appearance it retains to this day, leaving only traces of ridge and furrow to mark the work of the early medieval farmers.[25]

It should be emphasised, of course, that throughout the Middle Ages there were great variations in cultivation and land-use patterns from one part of the country to another. However, the country as a whole could be divided into two major agricultural zones either side of the line running from the Exe to the Tees. The first, that to the south and east, has been described as 'the lowland zone of mixed husbandry'; the second, that to the north and west of this line, is known as 'the pastoral zone'.[26] It was in the former area that the major changes from arable to pastoral farming occurred.

Finally, the later Middle Ages was marked by a massive abandonment of villages and hamlets, both in the marginal lands and in areas of relatively good soil, some 2,000 deserted medieval villages having been identified in England as a whole. The causes of this process, which began before the Black Death and continued throughout the later Middle Ages and beyond, are complex. They are partly to be found in a declining population, especially after 1350, and partly in a growing demand for wool in the fifteenth century which resulted in the conversion of much arable land to pasture, the eviction of tenants, and the desertion of villages.[27]

As in any period of economic recession, there were those individuals and families who were able to weather the storm and others who turned the general misfortune to their own advantage. They included great magnates, both of state and church, well-established families and, particularly, a new emerging class of yeoman farmers. The last group consisted of wealthier peasants who were able to take advantage of the economic situation of the late fourteenth and early fifteenth centuries by taking up, at low rents, leases of land being offered by the large landlords who had run into financial difficulties. Other emerging groups were the manorial officials, such as reeves and bailiffs who took up leases for the first time,[28] merchants and lawyers. Many of these 'new men' built up their landholdings and established local and national dynasties, dignifying their success by building substantial and sophisticated residences, particularly in the fifteenth century. These buildings, which are in sharp contrast to the relative poverty of much of the period, have left us with a rich heritage of late medieval architecture. It was during this period, too, that wealthy landowners created for themselves new large parks which differed from their early medieval predecessors in being 'amenity parks' rather than hunting grounds.

The towns of this period also witnessed considerable changes. Some of the older established towns, such as York and Lincoln, had their populations greatly reduced by the plagues of the second half of the fourteenth century and suffered from economic decline thereafter.[29] Others, both old and new towns, prospered from their close involvement with industry and trade. This was particularly true of cloth manufacture and trade which largely accounted for the growth during this period of such towns as Norwich, Bury St Edmunds and Exeter.

London, too, prospered from the wool trade and, by the end of the period, was far and away the largest town in the country.[30]

Fundamental changes also occurred in industrial development in the fourteenth and fifteenth centuries. Most significant was the growth of the textile industry, based on the gradual change from exporting raw wool to the Continent to exports of cloth. By 1450, this fundamental change had reached a turning-point,[31] and major producing regions, making cloth both for export and the home market, were the West Country (Devon, Somerset, Cornwall and Wiltshire), East Anglia and the West Riding of Yorkshire. The iron industry expanded, due partly to technological advances such as the increasing use of water power to work bellows, and the main centres of production were in the Weald, the Forest of Dean and the Cleveland Hills. Other important industrial activities included the mining of coal, especially in the Tyne valley, tin-mining in Cornwall, lead-smelting in the Peak District of Derbyshire and salt-making in Cheshire and Worcestershire.

By the beginning of the fourteenth century, the country's road system was already well established and, as today, it centred on London. Travel was on foot or on horseback, and goods were carried largely by horsedrawn carts and wagons, though bulky goods were sent by boat where possible. The road network and methods of travel were two of the few aspects of English medieval geography which were scarcely altered by the economic and social changes of the late fourteenth and fifteenth centuries.

Thus, the four hundred years of the Middle Ages in England were, by modern standards, a period of relatively slow change.[32] Nevertheless, considerable developments in the English landscape did occur and, through periods of prosperity and decline, some general trends are evident. The area of cultivated land increased, the woodland cover diminished, marshland and wasteland were reclaimed, parks were created, buildings were erected in the countryside, towns and villages were established and enlarged, industry was developed, and a network of roads and tracks was consolidated and extended. It is to a detailed examination of each of these major elements in the English medieval landscape that we shall now turn.

Notes

1. H.C. Darby (ed.), *A New Historical Geography of England before 1600* (Cambridge University Press, Cambridge, 1976), p. xii.

2. H.C. Darby, *et al.*, *The Domesday Geography of England* (Cambridge University Press, Cambridge, 1952-77), vols. I-VI; for a single volume work on the subject, see H.C. Darby, *Domesday England* (Cambridge University Press, Cambridge, 1977).

3. R. Lennard, *Rural England, 1086-1135* (OUP, Oxford, 1959), p. 393.

4. Darby, *A New Historical Geography of England before 1600*, p. 49.

5. D. Rackham, *Ancient Woodland* (Edward Arnold, London, 1980), pp. 126-7.

6. Darby, *A New Historical Geography of England before 1600*, p. 61.

7. Darby, *Domesday England*, p. 71.

8. R.A. Dodgshon, 'The Early Middle Ages, 1066-1350' in R.A. Dodgshon and R.A. Butlin (eds.), *An Historical Geography of England and Wales* (Academic Press, London, 1978), pp. 82-7.

9. See for example the Pelican History of England, vol. 3: D.M. Stenton, *English Society in the Early Middle Ages, 1066-1307* (Penguin, Harmondsworth, 1959).

10. Dodgshon and Butlin, *An Historical Geography of England and Wales*, pp. 87-8.

11. M.M Postan, *The Medieval Economy and Society* (Penguin, Harmondsworth, 1975), pp. 20-4.

12. Darby, *A New Historical Geography of England before 1600*, pp. 106-7.

13. A.R.H. Baker, 'Changes in the Later Middle Ages' in Darby, *A New Historical Geography of England before 1600*, p. 200.

14. For a detailed discussion of the various explanatory frameworks or models of the changes which took place in the later Middle Ages, see R.A. Butlin, 'The Late Middle Ages' in Dodgshon and Butlin, *An Historical Geography of England and Wales*, pp. 119-23.

15. Postan, *The Medieval Economy and Society*, p. 27.

16. C. Platt, *Medieval England* (Routledge and Kegan Paul, London, 1978), pp. 95-6.

17. J. Hatcher, *Plague, Population and the English Economy, 1348-50* (Macmillan, London, 1977), p. 25.

18. Postan, *The Medieval Economy and Society*, p. 43.

19. Baker, 'Changes in the Later Middle Ages', p. 195.

20. Platt, *Medieval England*, pp. 99, 102.

21. Baker, 'Changes in the Later Middle Ages', p. 197.

22. Ibid., p. 202.

23. A.R. Myers, *England in the Late Middle Ages, 1307-1536* (Penguin, Harmondsworth, 1963), p. 47.

24. L.M. Cantor, 'The Medieval Landscape of Staffordshire' in A.D. Phillips and B.J. Turton (eds.), *Environment, Man and Economic Change* (Longman, Harlow, 1975), p. 191.

25. C. Taylor, *Dorset, The Making of the English Landscape* (Hodder and Stoughton, London, 1970), p. 119.

26. J. Thirsk (ed.), *The Agrarian History of England and Wales*, Vol. IV, *1500-1640* (CUP, Cambridge, 1967), pp. 1-112.

27. M.W. Beresford and J.G. Hurst (eds.), *Deserted Medieval Villages* (Lutterworth Press, Guildford, 1971), pp. 12-14.

28. Platt, *Medieval England*, p. 133.

29. Baker, 'Changes in the Later Middle Ages', pp. 243-4.

30. Ibid., p. 246.

31. Ibid., p. 219.

32. Postan, *The Medieval Economy and Society*, p. 45.

2 MEDIEVAL FIELD SYSTEMS

Trevor Rowley

'The English landscape itself, to those who know how to read it aright, is the richest historical record we possess', so wrote W.G. Hoskins over a quarter of a century ago.[1] The concept of the English landscape as a palimpsest is now gradually being understood and accepted by historians, archaeologists and historical geographers. The landscape is a document onto which each generation inscribes its own writings and in the process removes some, but not all, of the writings of its predecessors.

In no area is this concept more valid than in the study of fields and field systems. Over much of central England the field system of regular hedged and fenced fields that we see today is a relatively recent product dating from the period of parliamentary enclosure in the eighteenth and nineteenth centuries. It is a landscape associated with straight roads, wide verges, and large isolated brick-built farm complexes. This landscape sits upon an earlier one, with which it only occasionally coincides. Even the casual observer will readily recognise extensive earthwork traces of an earlier agrarian system, a system that is most commonly associated with the Middle Ages – open-field agriculture (Figure 2.1). Its most frequent physical manifestation is in the form of fossilised ridges and furrows (Plate 1) which, when ploughed, formed the basic unit of medieval cultivation. This system itself was developed in part within an existing agrarian framework which may date back to before the Roman invasion of Britain.

Despite the introduction of 'prairie farming' and intensive ploughing campaigns over the last few years, traces of open-field agriculture in the form of earthwork ridge and furrow, soil marks and crop marks of ridge and furrow are ubiquitous in central England and surprisingly common in other parts of the country. This is largely because the enclosure of open fields was frequently associated with a change in land use from arable to pasture, which meant that the relicts of former ploughing activity were effectively frozen. In any case, it takes many years of determined cross-ploughing to eradicate completely the earthworks resulting from strip farming. These phenomena were so widespread that they still constitute the most extensive surviving remains of any earlier landscape (Plate 1). Despite its widespread occurrence

25

Figure 2.1: Earthwork traces of medieval ridge and furrow in the existing landscape

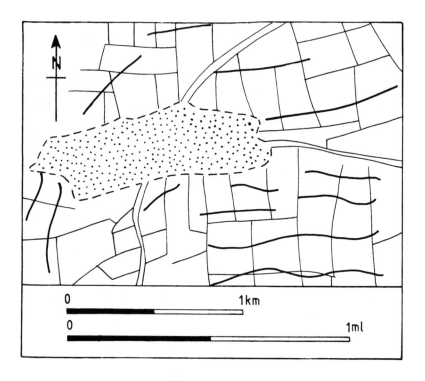

Source: C.C. Taylor, *Fields in the English Landscape* (Dent, London, 1975), p. 84.

ridge and furrow was not universal. For example, documentary evidence indicates that it was rarely formed in areas such as south Devon and the East Anglian Brecklands. In these areas it seems that strips were ploughed flat, without furrows, in order to retain moisture. Grass baulks or strip divisions may have been employed to define the strips. Certainly such baulks exist on the surviving open fields at Braunton in Devon (known locally as *landsherd*, Old English *landscearu*, boundary),[2] although even here they are rapidly disappearing through strip amalgamation.

In most parts of England, however, strip ridging was the normal method of ploughing, with the intervening furrows acting as property boundaries. Open-field agriculture survived in pockets throughout the

Plate 1: Ridge and furrow at Southam, Warwickshire

country well into the nineteenth century, and indeed modified open fields can still be found, most notably at Laxton (Nottinghamshire), where a virtually complete open-field system operates. On the Isle of Portland (Dorset), Braunton in Devon, Eakring in Nottinghamshire and the Isle of Axeholme elements of an open-field system still survive, and indeed similar interesting but incomplete examples are to be found in other parts of the country.

It should be stressed that the term 'open-field agriculture' is not synonymous with medieval agriculture (and indeed ridge and furrow does not have to be the product of open-field farming), although much if not most English medieval agriculture operated on an open-field basis. Difficulties arise in our understanding because the system which reached its zenith in the Middle Ages persisted in some areas for nearly four hundred years after what is generally thought of as the end of medieval England (1485). The dilemma which is created by this partial survival has made medieval agriculture difficult to analyse and a topic of considerable controversy over the years.[3] The great temptation for the historian and historical geographer examining medieval agriculture is to draw too heavily upon the extensive body of post-medieval evidence in order to reconstruct the agrarian system of the Middle Ages. The intrinsic dangers inherent in this approach are that topography and field systems which can be identified from post-medieval documents and maps cannot automatically be assumed to be medieval. There are rare examples of medieval maps which indicate in general terms the nature of the contemporary field systems, such as that of Boarstall in Buckinghamshire, 1444. Nevertheless, there is little authentic cartographic information available before 1600.

The problems of examining and identifying traces of true medieval agriculture are no easier for the archaeological field-worker, who rarely has the opportunity of examining unadulterated traces of medieval field systems. The archaeological remains of open-field agriculture date from the last occasion on which the fields were ploughed, which is often as late as the eighteenth or nineteenth century, and although they may provide an accurate facsimile of earlier systems they are not actually medieval. It is therefore easier to describe the legal, tenurial and operational aspects of medieval agriculture of the thirteenth and fourteenth centuries, than to explain its origins or its early development.[4] Similarly, we can be reasonably but not absolutely certain of its physical appearance in the thirteenth century, but not earlier. In the study of medieval field systems we have a true interface between the disciplines of history, geography and archaeology, with all the problems that are implied in the different nature and quality of the evidence and the manner in which it is treated.

Characteristics and Terminology of Open-Field Agriculture

Bearing in mind M.M. Postan's salutory warning that: 'It is even more dangerous to generalise about the organisation of medieval agriculture than its physical and demographic background',[5] we can be reasonably certain that the basic field system which operated throughout most of medieval England was an open-field system. Alternatively, this was known as the common-field, the two- or three-field system, strip farming or the Midland system. Because it is less contentious, carries fewer

Figure 2.2: The creation of ridge and furrow

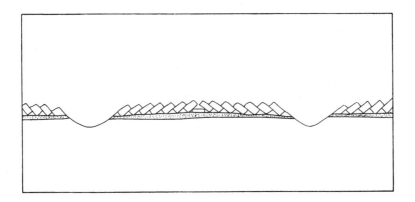

Source: C.C. Taylor, *Fields in the English Landscape* (Dent, London, 1975), p. 76.

overtones, and can be used more widely than any of the others, in this chapter the term 'open-field' will be used unless specifically discussing variant forms, including the infield–outfield or runrig system (see p. 36).

The authentic medieval open-field system consisted of four elements. First, the arable land was divided into strips or selions which were owned or tenanted by a number of people normally occupying a central settlement – a village or vill. Each strip was normally ploughed consistently as a discrete unit towards its centre (Figure 2.2). The result of this was to create high-backed ridges and furrows. The pattern of ploughing employed by the communal plough-teams resulted in giving them a characteristic S or C shape, known as the *aratral* (plough) curve. By no means was each ridge co-extensive with each strip or unit of landholding, but where studies have been carried out it can be seen that there was a marked degree of correspondence. Each landholder had scattered parcels or plough-strips of land distributed throughout the arable fields, intermingled with those of his neighbours.

Second, both arable and meadow land were pastured by the stock of the same farmers between harvest time and when the seed was sown. In most years approximately a third of the arable land within the open fields lay fallow in order to allow it to rest and be manured by the grazing animals (also providing the animals with valuable feed). This meant that on arable lands and, to a lesser extent, pasture and meadow, strict rules governing the nature of crop cultivation and the control of animals were necessary. Third, where there was pasture, waste or common land this was used for stock raising, again often involving strict control of the number and type of animals involved. Finally, the administration and implementation of this agricultural system had to be organised by a formal meeting of the farmers, normally in the form of a manorial court or a village assembly.[6]

One misconception that is commonly held about the open fields is that annually, or at least periodically, the strips were re-allocated among the landholders in order to ensure fair distribution of land of different quality within the community. Although there are many instances of redistribution of meadow land, and in some cases pasture, there is little evidence for the re-allocation of arable land on a regular basis. An exception to this generalisation appears to be Northumberland, where re-allotment of arable land was common in the sixteenth century, but not on an annual basis. In a number of villages the fields were divided into two halves and the strips re-allocated in order to give tenants land in one half or the other and thereby reduce the distances they walked to their parcels. It was also customary in parts of the northern counties to change the arable fields at intervals by putting the old plough land back to common pasture and taking a new field from the common.[7] There is, however, still controversy over the possibility of there having been open-field systems completely redesigned on one or more occasions in some instances.[8]

In its purest form the strip represented the individual unit of one man's holding. Although the system was not universally one of subsistence, it is clear that there was a considerable subsistence element involved in medieval agriculture by the thirteenth century over large parts of the country. It therefore follows that each farmer should have had a relatively evenly distributed land allotment, and not have had a disproportionate share of his strip under fallow in any one year. Also, each farmer should have had a share of ground on soils of varying quality. Both these features tended to break down, however, over a period of time, as strip amalgamation and swapping eventually resulted in considerable variations in the size of holdings (see Table 2.1).

The resulting pattern of widely dispersed strips was economically understandable, in so far as each farmer living alongside his neighbour in a nucleated settlement was centrally placed to deal with his scattered holdings.[9] It is noticeable that the enclosure of open fields in one form or another was apparently responsible for the breakdown of nucleated community. Striking examples of this process are provided in areas

Table 2.1: Approximate land divisions of Lower Heyford, 1604

	No. of strips	Acres	Roods	Perches	Percentage of total
Lampland	3	2	2	32	0.35
Thomas Guy	2	0	2	32	0.07
William Endall	4	3	2	19	0.48
Roger Wighton	5	2	0	18	0.28
Thomas Bruce	199	125	3	25	17.43
Barth Tipping	184	104	3	32	14.52
James Mynne	92	37	0	37	5.12
William Tredwell	162	45	0	27	6.22
Thomas King (e)	156	57	2	37	7.88
John Sheres	144	66	0	30	9.13
John Green	19	15	2	36	2.21
Richard Elkins	78	30	0	20	4.15
George Merry	126	60	2	35	8.44
John Merry	89	43	2	3	6.02
Gabriel Merry	119	104	1	21	14.38
Norton	92	22	2	9	3.11
Thomas Faulkoner	53	31	0	33	4.29
Simon Elfret	58	7	1	0	0.97
Parsonage	76	21	1	8	2.90

Source: Thomas Langdon, map of Lower Heyford, Corpus Christi College, Oxford (1604).

where fairly rapid eighteenth- and nineteenth-century parliamentary enclosure was particularly common. In Leicestershire, for instance, enclosure was fairly quickly followed by the rebuilding of farmsteads in the newly created farm units and not in the villages where they had previously been. This resulted in considerable village shrinkage in the middle decades of the nineteenth century.

The strips of the open fields were grouped into furlongs (*cultura*), 'flatts', 'shotts' or 'wongs' as they were known in some parts of the country. The furlong was not a unit of length, as it subsequently became, and furlongs could vary considerably in size even within a single manor. The groups of furlongs cumulatively formed fields (*campi*). Frequently there were three named fields of roughly equal size which in theory suited the rotational agrarian practice best, that is two-thirds arable, one-third fallow each year. Often the geography of the manor or territorial unit involved was such that a three-field system provided the most convenient and natural division of available land. Sometimes, natural geographical demarcations of drainage and relief or ancient land-use features, such as tracks or estates boundaries, automatically provided field boundaries.

Figure 2.3: The seventeenth-century four-field system on the Corpus Christi estate of Whitehall, Tackley, Oxfordshire

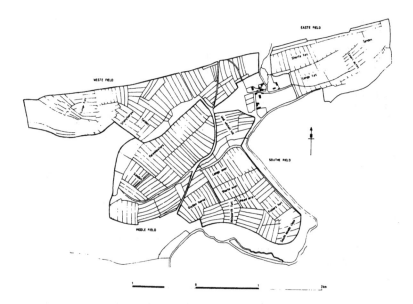

Source: Thomas Langdon, map of Tackley, Corpus Christi College, Oxford (1605).

It was not, however, essential to have three fields for this agricultural system to operate and much confusion about the development of the open fields has arisen because of the presence of only two fields or in some cases four, five or even more fields. The essence of the system was not the number of fields, but the rotational system operating within the field framework. This certainly did not have to be dictated by the number of named fields. Indeed it is quite clear that in cases where there were the conventional three fields operating, the rotation system did not rigidly follow the field system — the furlong, not the field was the rotational unit. In Clapham's words, 'crop rotation is independent of the lay-out of the fields'.[10] On the Corpus Christi estate of White-hall, Tackley, Oxfordshire an early seventeenth-century map and documentation shows a four-field system operating on perfectly regular rotational lines, with the furlongs acting as the unit of rotation (Figure 2.3).

There has been much argument about the nature of the divisions between the strips. Originally it was thought that *balks* (or baulks), consisting of narrow strips of grass, divided the strips from one another. Although in parts of the country these did exist, it would appear that normally no such divisions were to be found. Alternatively, boundary or mere stones, or even posts or sticks, have been suggested and, although evidence for these occasionally appears, they were certainly not a general characteristic of the system. Trackways which served the furlong were also known as balks in some cases and later survive as green ways.

Within the open-field system there were marked regional variations. The reasons for this were not ethnic as suggested by early scholars, but historical and geographical. It can no longer be accepted that the open-field system was a product of the Anglo-Saxon migration. There is no evidence whatsoever to suggest that there was a major change in agrarian organisation between the fifth and eighth centuries AD (see p. 41). What evidence there is indicates the survival of Romano-British fields and boundaries into the post-Roman period. When these were adapted to the conventional medieval form of field system, the form of fields already operating must have heavily influenced the layout of the new field systems. For instance, rectangular Celtic or Romano-British fields were subdivided into blocks or strips and the resulting furlongs were of a different size and shape from those developed from the subdivisions of smaller rectangular or long fields (Figure 2.4). Second, the varied physical environment of England, in terms of climate, soils, slope and aspect, resulted in different agricultural regimes which would have been reflected in different field systems. Added to this, differences

in land tenure and inheritance customs meant that there could be an almost infinite variety of field forms. This is quite apart from those parts of the country where open fields never seem to have operated, such as Cornwall.

Another fundamental and much discussed aspect of medieval agriculture was the importance of population to field systems. Where and when there was considerable population pressure, this resulted in the maximum use of land for arable purposes, culminating in some cases in the complete utilisation of the available land area. In other areas, extensive tracts of land remained waste and common; in the western upland areas, for example, waste land and marginal agricultural land remained widely available throughout the Middle Ages. In the lowland areas where, especially in the twelfth and thirteenth centuries, there was great population growth, grazing and meadow land became increasingly under pressure. For example, an examination of the extent of ridge and furrow in parts of Warwickshire and Northamptonshire in the thirteenth and early fourteenth centuries clearly shows the total use of land for arable (see Plate 1).

In the Cotswolds and other hilly parts of central England this is reflected in the incorporation of steep slopes into the strip system, by way of terracing; these are commonly known as strip lynchets (see p. 52). It is in just these central clayland areas of England that when the population fell during the fourteenth century we see the most dramatic evidence of both arable and settlement abandonment (Plate 7). Large tracts of former ridged arable were turned over to pasture, resulting in some of the finest medieval earthworks in England.[11]

The variations that do occur often coincide with areas where the regular pattern of nucleated villages or hamlets did not exist in the Middle Ages. If we accept that settlement patterns and field systems are inextricably linked, it is often impossible to discover which preceded the other. We can, however, see that both are linked to the physical geography and inheritance customs of a region.

Emphasis has already been placed upon the considerable degree of regional variety to be found in open-field farming. In south-east England, for example, a system of inheritance known as partible inheritance, where land was divided equally between all male heirs, resulted in a curious pattern of small blocks of open or strip fields not always cultivated or grazed in common, mixed up with areas of enclosed fields. Some of the latter were apparently cultivated in strips for cropping purposes and later sold, leased or inherited as individual strip holdings.[12]

Another interesting variation, which may throw some light on the origins of the system, is to be found in the Holderness area in the low-lying region to the east of the Yorkshire Wolds. The townships normally had only two arable fields. Many strips within the fields were of great length, often extending from one field boundary to another, a distance of over one mile in some townships. Strips lay parallel throughout the greater part of a field and furlongs, where these occurred, were few in number within any one field, and were usually functional rather than physical divisions within the fields.[13] Harvey has suggested that these field systems were reorganised, perhaps along with their associated settlements, during the late Saxon or early Norman period.

In the systems of the far west – the uplands of west Somerset, mid-Devon and the Welsh Marches – quite extensive tracts of countryside were characterised by open-field systems, which had a form of history different from both those of the Midland zone and those of the uplands. If we take an example from east Devon, which lies between the western boundary of the Midland zone and the Exe valley some twenty miles to the west, in the early Middle Ages the villages of the lowlands had field systems with some similarities to those of the Midlands. Many of them were surrounded by extensive tracts of arable land, the arable lay in strips within units designated as fields and furlongs, but the diagnostic attributes of the Midland system were missing; the compact fallow field too was absent, and so too was the practice of village-wide common pasturing on the uncropped arable. The arable land of each settlement did not comprise two or three great fields, each subdivided into furlongs, but rather a multiplicity of units, some called fields, some furlongs. The earliest evidence for east Devon reveals the presence of these irregular multi-field systems and no hint of more regular two- or three-field arrangements in the thirteenth century. By the sixteenth century most of this arable had been enclosed and much of it turned over to pasture. The agrarian history of the region is not one of decay of the two- or three-field systems but the transformation of an indigenous and flexible multi-field arrangement. We must assume that such variations on the accepted system were common throughout western areas.[14]

Another important variation on three-field systems was the 'runrig' or 'infield–outfield' system which appears to have operated on poorer ground in north and western England. This consisted of one field, the 'infield' which was closest to the vill or farmstead, and under perpetual cultivation; and a second field, the 'outfield', which was divided into two, one part being cultivated for several years and then grassed for a

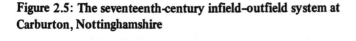

Figure 2.5: The seventeenth-century infield–outfield system at Carburton, Nottinghamshire

Source: Nottinghamshire County Record Office, SP 14/83/80.

few years, the other part being very largely uncultivated. It has been demonstrated that the 'infield–outfield' system could operate in a wide variety of different ways. Indeed, some scholars have argued that the 'infield–outfield' system was the foundation from which the two- and three-field system evolved. There are instances in Shropshire where it is possible to demonstrate such a development,[15] but it would be rash to postulate this as universal.

In some cases, probably in areas where population pressure was never a real problem, the infield–outfield system survived without substantial change throughout the Middle Ages. At Carburton in the forest area of Nottinghamshire an infield-outfield system survived into the early seventeenth century, when it was recorded in cartographic form. In the area immediately adjacent to the town there were regular open-field strips. There is, however, a large area known as the Brecks which was common land, available when required for cultivation but not a permanent part of the ploughlands[16] (Figure 2.5).

Enclosed Fields

By 1300 there were various forms of open-field farming over much of England; this type of cultivation was, however, not employed everywhere. There were large sections of the country where no strip fields existed and agriculture was based entirely on enclosed fields which were bounded by permanent hedges, walls or banks. These fields varied greatly in shape, size and purpose but generally they developed mainly in localities with a relatively low population, where the isolated farmstead or hamlet was more common than the village, and where pastoral farming was more important than arable.

In parts of the far west country the patchwork of enclosed fields that operated during the Middle Ages was directly descended from prehistoric field systems. In these areas there had been no need to change tenurial organisation or farming practice. Recent work by Glanville Jones has demonstrated the survival of Celtic agrarian practice in western Britain into the Middle Ages.[17]

Reference has already been made to the enclosed fields which could appear on the edge of the open fields on land that had earlier been woodland or waste, and throughout the Middle Ages there were manors even in central England where all or sections of the strip fields were being hedged. In the west, particularly as the population fell dramatically during the later fourteenth and fifteenth centuries, former open fields were turned over to pasture; so much so in fact that in counties such as Herefordshire and Shropshire by the time the first cartographic surveys were made only very small areas of open fields survived. In these areas the absence of parliamentary enclosure deceived many scholars into believing that open fields had never operated. The examination of the documentary evidence and the identification of ridge and furrow, however, show clearly that strip farming was associated with virtually every nucleated settlement in the thirteenth century.

From about 1300 onwards, the process of enclosure by exchange of strips by tenants and the amalgamation in small fields became a regular feature of the landscape. This process was often associated with village shrinkage or even desertion. In the case of the Hundred of Ford in Shropshire, out of 70 Domesday hamlets, nine had been completely deserted and 33 shrunk to single farms. Indeed only twelve of the settlements survived as nucleated settlements. Much of this shrinkage occurred during the later Middle Ages when a large proportion of the land changed from arable to pasture.[18] It is quite clear from the documentation that each of these small nucleated settlements or hamlets had its own open-field system. Much of this enclosure was undertaken by a private agreement, perhaps without any documentation, even at the time. In other cases, the repeated stricture of the manor courts to tenants telling them to pull down fences or newly planted hedges attests to the gradual attrition of these small open-field systems.

Often small bundles of strips were enclosed together in irregularly shaped fields, frequently in the immediate vicinity of the settlement. One of the features of this type of early enclosure is that the field boundaries often respect the boundaries within the former open fields; that is they conform to former furlong boundaries, unlike later parliamentary enclosure where the enclosure commissioners completely ignore former land units. Thus, in some cases the impression is given that because the surviving ridge and furrow is so well accommodated within the later field systems the strips were ploughed within the enclosed field system. In other cases where ridge and furrow was not formed, no surviving trace of these earlier open-field systems survives.

The enclosure of areas of former waste and woodland was associated with the growing population and a need to increase the area of agricultural production. From the documentary evidence it appears to have reached its height in the thirteenth century, when it is clear that there was extensive assarting. For example, in the parish of Whiteparish in south Wiltshire where there were relatively small areas of old forest, the fields are today highly irregular and have obviously been cut out of the former woodland, but the actual farm boundaries are still visible as relatively straight lines. From this one can see exactly how each farmstead was set upon the forest edge and the land behind it cleared piecemeal into long strips up to the cleared land of adjacent farms. Some of the actual clearances are recorded in documents, as in one of 1270 when 14 acres of assarts were enclosed by 'a dyke and a hedge'.[19]

The recognition of areas of woodland later cleared by medieval farmers can be taken even further by employing botanical techniques. Recent research by historical ecologists has lead to the identification of relict woodland hedges. These often appear to be ordinary hedges but are normally botanically rich. They contain plants such as dog's mercury, bluebell and wood anemone, all of which are slow colonisers and therefore indicators of old woodland. The occurrence of hedges with these species can often point to the existence of ancient woodlands cleared in the medieval period when no other evidence survives. Another well-established botanical technique which is of great value in understanding medieval enclosed fields is hedgerow dating; this involves counting the number of shrub or tree species along a 30-metre length of hedge. The average number of species in a 30-metre section equals the approximate date of the hedge in hundreds of years. Although this technique can not be universally applied it can be of considerable value in dating the creation of fields.[20]

Forests and wastes were not the only areas to be reclaimed for new arable land in the Middle Ages. In the uplands of northern England, along the Welsh borders, as well as on the moors of south-west England, reclamation was common. In the north many of the great Cistercian abbeys reclaimed vast areas of land and turned them into large arable farms. Elsewhere in Yorkshire much piecemeal assarting took place, not all of which was for arable, much was for sheep. Similarly, all round the edges of Dartmoor reclamation went on apace. The assarted farmsteads around the edge of the moors are today still surrounded by small irregularly shaped fields bounded by stone walls which were first created in the Middle Ages. In the Welsh border large areas of land were enclosed and newly established farms, often moated, are located from this period.

Work also continued on the reclamation of fenland and marshland in the west of the Somerset Levels; drainage started by the twelfth century and thirteenth century was mostly piecemeal in character. In eastern England landowners were engaged in the extension of arable across the level silt-lands around the Wash. Some of the new land was incorporated into the existing strip fields, but much was enclosed directly into small fields. This process certainly started before the eleventh century and continued throughout the next two hundred years. Co-operation was required to carry out this work often by groups of villagers, the basic requirement being the construction of massive banks running parallel to the sea. Behind these a new level was drained, turned over to pasture and in some cases cultivated. In a survey of 1251 at Little Port, Cambridge; 60 new tenants were holding nearly 200 hectares of new land and this is reflected in the field pattern on the ground today.[21] In the case of Rennington in Northumberland, a map of 1624 shows the west field as being disproportionately large and we can gather from this that it had been enlarged by assarting from Rennington Common.[22]

Origins and Development

Although scholars agree about most aspects of the nature and operation of the 'mature' open-field system as it is termed for the thirteenth and early fourteenth centuries, no such agreement exists over the way in which the system originated or over its early development. The study of medieval field systems dates back to the late nineteenth century when scholars such as Nasse,[23] Seebohm,[24] Vinogradoff[25] and Maitland[26] were already deeply involved in dispute about the nature and origins of open-field agriculture.

Although Seebohm was intent upon tracing English rural institutions back to Roman or pre-Roman roots, the general consensus was that open-field agriculture was a product of Anglo-Saxon settlement in the immediate post-Roman period. This view was amplified by an American scholar H.L. Gray, who was the first person to scrutinise carefully a large sample of evidence relating to English field systems, and who produced the work which has largely dominated thinking on the subject this century – *English Field Systems*.[27] Gray argued that the classic open-field system, or 'Midland system' as he termed it, was imported into this country, along with the nucleated village, by Germanic settlers. He also contended that the alternative systems found in East Anglia and Kent, for instance, could be explained in ethnic terms, resulting from different tribal groups bringing in their different agrarian traditions.

It was not until 1938 when the Orwins wrote their seminal book on medieval agriculture, *The Open Fields*,[28] that any significant development in thinking took place. Whilst the Orwins did not refute that the system had cultural origins, they were much more concerned with its practicalities. They portrayed groups of settlers who, in order to survive, had to co-operate in the process of bringing woodland and waste into cultivation. The resulting arable land was divided into strips and managed according to common rules. The Orwins believed that biennial rotation, with a fallow every second year, was a natural practice for any group of settled farmers and that there would originally have been two fields; subsequently the two fields were made into three, or a new field was added by further clearance, a process known as *assarting* (which literally means 'grubbing up' in Old French). This model had the advantage of being eminently logical, while not placing too much emphasis upon the cultural differences of the incoming settlers.

In retrospect, it can be seen that the Orwin model had two weaknesses. First, it assumed (quite naturally for the time) that the incoming settlers of the post-Roman period completely replaced the indigenous population in central and eastern England, in other words that there was no continuity; and second, that much of England still lay under a mantle of primeval woodland waiting to be tamed and cultivated. Neither of these assumptions is acceptable today; archaeological work over the past two decades has demonstrated that there was a considerable degree of survival of earlier features in post-Roman England, particularly in the form of settlement and estates. It is also reasonable to assume that Romano-British agricultural practices survived over considerable parts of the country. It is also now clear that most of lowland England had long since been cleared of its natural woodland and that the landscape had been carefully managed, probably since the Bronze Age. In other words, by the fifth century AD, the English landscape already contained well-established field systems worked under well-established agrarian practices; it was not a clean slate on which the Anglo-Saxons could, even if they wanted to, write a completely new story.

Nevertheless, early medieval documentation demonstrates that assarting was commonplace. This must mean that land which once had been cultivated had degenerated to scrubland and was being re-cleared as a result of population pressure or, more likely, in parts of the country there still remained pockets of marginal land available for clearing. It has been argued that the furlong was a composite unit, representing blocks or arable cleared by communal effort from common, waste, or perhaps pasture, and divided between the participants responsible for the clearing. The beauty of this argument is that it would explain the irregular nature of the furlongs, the reason why they change direction, presenting the patchwork-quilt appearance so commonly seen in aerial photographs and early estate maps. The alternative argument — that they represent subdivided Celtic fields — also has its attractions, and the truth must lie somewhere between the two.

An examination of glebelands (the land apportioned to the rector along with the living of a parish) within the open fields in some areas indicates that there was a regular pattern of apportionment. Glebe strips, unlike other holdings, tended to remain in the same pattern over many generations, thereby providing a fossilised form of land distribution within the field system. For example, at Langtoft, East Riding, the glebe strips were the 11th, 12th, 26th, 27th, 30th and 31st in each furlong as late as the eighteenth century, and at Great Givendale, East Riding, in 1684 the parson's strip was always the end one in a furlong, with the lord of the manor's lying next to it. Beresford has persuasively argued that regularity of this type was not accidental and resulted from successive allocations of newly cultivated or divided lands.[29] A glance at the distribution of parsonage land within the strip system of Lower Heyford (Oxfordshire) clearly demonstrates a fairly regular arrangement compared to the comparable-sized holding of another tenant (Figure 2.6).

This is an appropriate place to discuss the degree of regularisation in medieval settlement and fields. It is now quite clear that there was a considerable amount of village creation, or perhaps more properly re-creation in the early Middle Ages. The regular form of many village plans indicates their designed origins; we have seen something of this process in Holderness, Yorkshire (p. 36). In the case of East Witton in Wensleydale, Beresford has suggested that the monks of Jervaulx Abbey who acquired control of the township by the late thirteenth century were responsible for the development of a new village on the edge of a street green, designed to provide a setting for a Monday market and Michaelmas fair first held in 1307. The almost geometrical layout of the village even today suggests that there was a high degree of planning in its design.

It has been suggested that such a redesign could be reflected in the division of strips within the open fields. This division is commonly known as *Solskifte*. In this system the order of strips within the furlong reflects the order in which the houses lay along the village street. This was derived from some original layout by which the strips within each freshly colonised furlong were distributed in the order which houses lay along the village street or around the green, although it is impossible to determine how far back in time distribution of this sort took place. It has been suggested that the distribution of strips within the open-field system at Wharram Percy, Yorkshire, conformed to this system. The order of arrangement of tofts and crofts along the principal street of Wharram Percy appears to indicate a *Solskifte* arrangement which numbered off its houses in the order that the sun moved over them.[30]

A third weakness in the Orwins' argument, however, is one which is shared by the majority of field studies, that is the over-reliance on

Figure 2.6: The distribution of parsonage land, Lower Heyford, Oxfordshire

Strips held by Gabriel Marry

Strips held by The Parsonage

Source: Thomas Langdon, map of Lower Heyford, Corpus Christi College, Oxford (1604).

post-medieval documentation to reconstruct medieval and earlier agrarian topography. Nevertheless, the study of Laxton, which forms the second portion of the book, remains the most thorough analysis of any open-field system in England.

The next major change of attitude on the part of historians came in the mid-1960s when Thirsk proposed a radically new model for the origins of the system.[31] According to this the system originated through the refashioning of earlier, irregular multi-field arrangements, in which a large proportion of a community's arable had already become divided into strips. However, these earlier field systems lacked the other features of the Midland system and, instead exhibited a 'puzzling appearance of discorded cultivation,' each township having 'numerous fields, not apparently arranged in any orderly groups'. As far as pasturing was concerned, agreements were made 'between neighbours possessing intermingling or adjoining land'. But the great compact fallow field was absent in both concept and practice. The Thirsk model therefore states that the principal physical ingredient of the open fields − the strips − came before the common agricultural system. Arguing largely from German examples, she claims that the strips were created as a result of population pressure linked to the development of nucleated villages, which brought about the subdivision of existing fields into smaller and smaller units. 'A multitude of German examples can be cited of townships consisting at one stage of large undivided rectangular fields, which became subdivided into hundreds of strips in two centuries or less.' The fields were divided into long strips, rather than small rectangular fields, simply because it was more convenient for cultivation by the plough.

The stimulus which encouraged remodelling of systems of this kind was a growing need, as communities grew and waste diminished, to use fallow as efficiently as possible, a need which brought them to a point at which it was necessary to introduce a system 'in which all tenants shared common rights in all fields'. In the interests of the viability of individual holdings, a 're-distribution of land was necessary in order to facilitate the introduction of new common-field regulations'. This remodelling of field systems might be achieved quite rapidly or through exchanges made over a number of years, processes which took place 'at different times in different parts of the kingdom'. Thirsk concludes that, in general, 'rights of grazing over arable land were still being shared by neighbours in the twelfth century, but before the middle of the thirteenth century there were villages in which all tenants shared in common rights in all fields', and that 'we can point to the twelfth and first half of the thirteenth centuries as possibly the crucial ones in the development of the first common-field system'.

The Thirsk model has been spelt out in detail because it has a number of very attractive features. It finally dispenses with the concept of ethnic origins of the open fields, replacing it with an economic explanation — an explanation that can be clearly demonstrated in other geographical contexts, for example, over large parts of northern and central Europe. The rejection of the ethnic origins also incidentally releases the creation of strips from being restricted to a particular chronological period; instead, the process could take place at any time, whenever and wherever there was sufficient population pressure. It also provides a strong logical reason for the change from square and rectangular 'Celtic' fields to the strip fields, which in landscape terms is the fundamental change that occurred. Strips appear to have played a relatively insignificant role in earlier field systems in England.

Needless to say, the new model produced intense interest, not to mention dissent. The strongest reaction has come from Titow who has basically reiterated H.L. Gray's more gradual colonising approach.[32] Broadly speaking, however, the Thirsk explanation of the mechanics of the creation of the system has been accepted, while its dating to the thirteenth century has met with considerable scepticism. The debate continues and has recently been joined by H.S.A. Fox, who in an extremely thoughtful contribution argues for a flexible approach, but for a largely late Saxon date for the introduction both of strip farming and for common agricultural practices.[33]

Fieldwork, Archaeology and the Open Fields

Only recently has much serious attention been paid to the surviving physical traces of open-field agriculture. For the most part this takes the form of roughly parallel ridges and furrows arranged in blocks (see Figure 2.6 and Plate 1). The topic of ridge and furrow was for some time the subject of fruitless argument over its origins.[34] The sheer quantity of ridge and furrow tended to blind scholars to its value as a source for investigation into the nature of medieval agriculture. Indeed, some writers implied that the very fact of its existence as a common element in the landscape devalued it as a serious source for investigation.[35] Pioneer work by Mead in Buckinghamshire indicated that there was a high degree of correlation between ridge and furrow and the individual strips shown on early estate maps,[36] and therefore presumably with medieval land units. Subsequent research has shown that this is far from being universally applicable, but that generally speaking ridges do tend to correspond to former land-holding units over much of the country.[37]

We must now turn to the nature and function of ridge and furrow and the other surviving features of open-field agriculture. Ridge and furrow consists of long narrow ridges of soil, lying parallel to each other and usually arranged in roughly rectangular blocks, separated by depressions or furrows. It is produced by the action of ploughing with a 'heavy' plough, that is a plough capable of turning over the sod and which, accordingly, has a share, coulter and mouldboard of some form. Therefore ridge and furrow can date from any period after the introduction of such a plough and is not necessarily medieval in origin. Nor does it necessarily have to be associated with common open-strip fields.

Ridge and furrow is produced by first ploughing a normal furrow across the field. Then, on the second or return run, a furrow is cut closely parallel to the first and the sod turned inwards to meet the one cut from the original furrow. The third run is then made next to the first, with the sod turned over the original one. Thereafter the process is continued back and forth across the field. Obviously such initial ploughing does not build up earthen ridges by itself. Indeed, constant ploughing over a long period of time, following the exact form of the original strip, is required to enable the ridges to develop to any height. There appears to be a correlation between the height and breadth of ridge and furrow and the nature of the soil. The heavier the soil the more capable it is of holding large ridges. At its largest it can be up to 1.75 metres from the depth of the furrow to the height of the ridge, as on the heavy Midland clays. Similarly, the width of a ridge can vary from one to five metres. The size of the ridges is affected both by the capacity of the soil and also by the period of ploughing. It, therefore, follows that the very largest ridges tend to be those on clays and marls which have been ploughed in the same fashion over many decades.

Despite the fact that ridge and furrow can be produced at any time, one has only to examine areas of it on the ground to recognise a relationship between the shapes and layout of the ridges and furrows and the known pattern of open-field strips. The individual blocks of ridges can be seen as equating with the blocks of strip, the furlongs, and the varied arrangement of the end-on and interlocking strip furlongs is repeated exactly in the pattern of ridge and furrow furlongs. Thus, there can be no doubt that at least part of the ridge and furrow still perserved over many parts of England is closely connected with the former medieval fields. Nevertheless, it must be stressed that their form was imposed during the last time they were ploughed in that way; strictly speaking surviving ridge and furrow is not medieval at all but relatively modern, dating from the latter half of the eighteenth or earlier part of the nineteenth century.

We can be reasonably sure that medieval farmers created ridges as early as the thirteenth century, for at this time Walter of Henley advocated the ridging of fields.[38] More definite and earlier proof has come from archaeological investigation. However, authentic examples of demonstrable medieval ridge and furrow are rare, partly because the ridges have to be stratified, that is to say sealed by a clearly datable layer. The opportunities for excavating field systems in such contexts are also rare, and the techniques required are of a most painstaking quality in order to locate the undulations in a buried ground surface. Such evidence has, however, been found at Hen Domen, Montgomery, just over the Welsh border from England. Here a late eleventh-century motte-and-bailey castle appears to have been built over shallow ridge and furrow. The removal of a portion of the castle bailey and the location of the pre-castle ground surface confirmed that the ridge seen in earthwork form outside the castle did indeed originally continue in the area where the castle was built.[39]

At Bordesley Abbey at Redditch, Worcestershire, the excavation of a twelfth-century monastic precinct bank has provided evidence of earlier ridge and furrow sealed underneath.[40] It seems probable that these features belonged to a field system of the village of Bordesley which was deserted at the time of the Abbey's creation, about 1140. Both Hen Domen and Bordesley demonstrate the existence of ridge and furrow in the period 1000–1200, but are unable to provide further clues on dating. At Wharram Percy in Yorkshire, the excavation of a deserted-village boundary bank suggests the creation of a field boundary for open-field strips of the same Saxon–Norman date.[41] Evidence of a less precise nature from the deserted village of Upton in Gloucestershire, suggests a similar dating for arable, but not ridge and furrow.[42] In a few cases earlier material stratified beneath ridge and furrow suggests that the ridging did not move after its creation. Romano-British floor surfaces were found intact beneath the open-field ridges at Chesterton in Warwickshire and Humberstone in Leicestershire.[43]

Other early examples are provided by earthwork evidence, as at Bentley Grange in Yorkshire, where the upcast of bell-pit coal mines dated to the twelfth century clearly sits on top of narrow ridge and

Figure 2.7: Braham Farm, Cambridgeshire

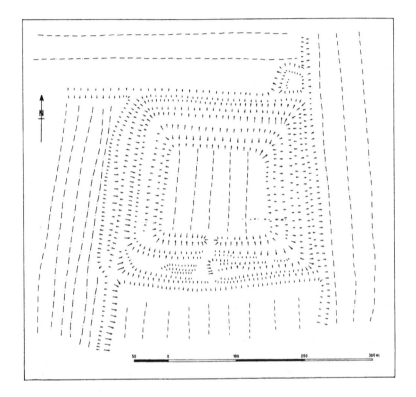

Source: C.C. Taylor, *Fieldwork in Modern Archaeology* (Batsford, London, 1974), p. 60.

furrow.[44] At Braham Farm, Cambridgeshire, an enclosure of almost certain medieval date overlies ridge and furrow (Figure 2.7).[45] Many aerial photographs of deserted medieval villages show how, during a period of expansion, existing tofts are extended onto or new tofts created over ridge and furrow. In the case of Barton Blount in Derbyshire, aerial photographs show clearly that the village has expanded northwards over earlier ridge and furrow.[46] Indeed, in a number of cases it appears as if a whole medieval settlement has been superimposed on a ridge-field system. At Newbold Grounds, Northamptonshire, an aerial photograph clearly demonstrates that earthworks of a medieval settlement lie over former ridged arable.[47] In this instance the place-name is post-Norman Conquest, which probably confirms settlement expansion during a period of population pressure during the twelfth and thirteenth centuries. It has also been suggested that the burgage tenement plots of some planned medieval towns were laid out over open-field strips. The long sinuous burgage plots to be seen at Thame in Oxfordshire and Stratford-upon-Avon in Warwickshire, would appear to fall into this category.[48]

The analysis of ridge and furrow is of particular value when applied over wide areas. It is usually impossible, using sixteenth-century or later maps showing open fields, to know exactly what the maximum extent of the medieval fields was in many places. This is because in later and post-medieval times there were changes to field systems throughout the country. Contraction and even abandonment took place long before the earliest maps were made, so that the identification and mapping of ridge and furrow can indicate where cultivation took place if there is no supporting cartographic or documentary evidence. Field surveys of ridge and furrow are also valuable in areas where it is known that there has been no ploughing for some time. Parkland areas created in the late medieval period are important in this respect. A particularly good example of this kind of work is the recognition of the very slight ridge and furrow which occurs on the chalk downlands of southern England.[49] Until the Second World War, this type of ridge and furrow was widespread, and although modern destruction has now obliterated all but a few fragments, it remains of considerable interest. First, the ridges are often almost imperceptible and only a few centimetres high, indicative that the ploughing that formed them was of relatively short duration. Second, the ridge and furrow often lies on top of older prehistoric and Romano-British fields, whose remains can be clearly identified under the ridges. This means that the type of ploughing which provided the ridge and furrow was relatively short-lived, for a long period of cultivation would have destroyed all traces of the underlying fields. Instead, in many cases, the earlier fields are still almost complete, with only their sides slightly ploughed down. Therefore, this ridge and furrow represents downland taken into cultivation for only a short period, at a time of acute land shortage, presumably in the thirteenth century.

Most of this ridge and furrow is undated, though in some places it is overlain by later ploughing of the eighteenth or nineteenth centuries proving that it is of some antiquity. One area of such ridge and furrow, on Fyfield Down, Wiltshire, has been dated because of its association with a small farmstead. This indicated that the temporary ploughing here was of late thirteenth- or early fourteenth-century date. This is not unexpected, for its falls into the period of the known maximum extent of arable land all over the country, when the population was probably higher than it was to be for the next few hundred years. If the date for the Fyfield Down fields is acceptable for similar areas elsewhere, then this ridge and furrow represents the final push of the medieval farmers into the marginal lands at a time of desperate land hunger. However, these areas were obviously soon abandoned and without the identification of this ridge and furrow type, an important part of the medieval agricultural history of the area would have been lost. Recently, very slight ridge and furrow of this type has also been identified in large quantities in the Welsh borderland, often in areas that have been under pasture for many centuries. Although it is still not absolutely certain why it came into being, it seems likely that it was the product of large-scale ploughing, again perhaps linked to population pressure in the twelfth and thirteenth centuries.

One of the incidental results of the massive extension of the ploughing up of pasture land during the past decade has been to produce new evidence for the laying out of strip systems in the form of soil marks. On the Berkshire Downs, for instance, a number of recent aerial photographs shows clearly how the strips were established within an already existing framework of rectangular fields,[50] indicating that in parts of the country, at least, the strips represented the subdivision of already existing field systems. This phenomenon has been observed elsewhere in the Midlands,[51] and cannot be dismissed as an isolated example. The change-over cannot be dated, but the fact that it took place within an extant system of land units tends to support a theory about the mechanism of strip creation.

None the less, despite its widespread occurrence, no really satisfactory explanation has been offered to account for the purpose and use of this technique of ploughing in ridges. The most usual explanation is that it is a response to drainage problems on heavy soils. It has been suggested that medieval farmers used the techniques of ridged ploughing to provide a way of removing water from the fields by allowing it to run along the furrows and, indeed, support for this argument may be seen in the actual arrangement of ridge and furrow. For example, an examination of all the ridge and furrow in 37 parishes of western Cambridgeshire and 120 parishes in Northamptonshire has shown that on almost all slopes of more than five degrees the ridge and furrow runs at right angles, or nearly so, to the contours.[52] In addition, where there are natural water courses such as streams and rivers, the ridge and furrow nearly always runs at right angles to them, thus facilitating the running away of water. Certainly contemporary accounts suggest that medieval farmers believed that ridging facilitated drainage. At Shipton in Shropshire, a court order of 1553 directed that 'everyone does plow the land so that the rainwater may run thereof'.

On the other hand, the existence of ridge and furrow on chalk downland, limestone hills and dry gravel terraces, where no drainage problem exists, has cast doubt on the drainage theory. In the upper Thames Valley, for instance, where there has been intensive aerial survey for prehistoric cropmarks over the past fifty years, ploughed-out ridge and furrow forms the most common crop-market feature, and is found virtually everywhere overlying earlier features. An alternative theory put forward to explain ridging, but one which has engendered little support, is that it increased the surface area of the field, and so enlarged the amount of land available, if only by a small proportion.[53] The true answer may be that the technique of ridging was originally developed as a way of draining heavy land, but that it subsequently became the normal way of ploughing, regardless of environment.

Apart from ridge and furrow, there are also other remains of medieval ploughing still visible in the modern landscape. These take the form of *strip lynchets*, a confusing name that has been given to the long parallel terraces which are commonly seen on steep hillsides. They occur widely in southern and south-eastern England, Gloucestershire and the uplands of northern England, but also appear less obviously in areas such as Essex, Hertfordshire, Cambridgeshire, Staffordshire, Cornwall, Herefordshire and in many other places. They have excited much interest in the past because they are often of considerable size and complexity. Explanations for their occurrence have ranged from describing them as terraces cut by river action to medieval vineyards, but in fact they are nothing more than the extension of the normal medieval open fields on to steep ground at a time when flatter, more easily worked land was in short supply.

A Postscript on Ridge and Furrow

The debate on the method of creation and function of ridge and furrow has recently been re-opened with spirit. Some scholars now claim that wide high-backed ridge and furrow, at least, could not have been built up by ploughing. They argue that it is physically impossible to create the characteristic humped profile by persistent ploughing and that such ridge and furrow must have been dug by hand in the first instance. Furthermore, the making of the aratral curve would have been extremely difficult and called for a totally unnatural method of ploughing. The argument continues that the creation of ridge and furrow must be seen in the light of the great medieval convention of earth-moving, found in castles, moats and fish-ponds. It is, of course, not contended that, once built, ridge and furrow was not subsequently ploughed. Even here, however, considerable doubt has been cast upon the conventional belief that all ridge and furrow was necessarily arable. The debate continues.

Thus, the landscape associated with medieval farming was not a uniform one, but varied considerably in both time and space. Elements of this landscape survived in use until the nineteenth century and, substantial traces of it survive in boundary and earthwork form today. It is a salutary comment on the increasing speed of our ability to change our environment that the processes which are now destroying the vestiges of medieval agriculture are taking with them the subsequent landscape as well. The removal of features associated with former agricultural systems is now so thorough that in parts of England it is effectively obliterating the former rural landscape altogether.

Notes

1. W.G. Hoskins, *The Making of the English Landscape* (Hodder and Stoughton, London, 1955), p. 14.

2. M.W. Beresford and J.K.S. St Joseph, *Medieval England: An Aerial Survey*, 2nd edn (Cambridge University Press, Cambridge, 1979), pp. 43-4.

3. The principal papers in the most recent controversy, J. Thirsk, 'The Common Fields', *Past and Present*, no. 29 (December 1964); J.Z. Titow, 'Medieval England and the Open-Field System', *Past and Present*, no. 32 (December 1965); and J. Thirsk, 'The Origin of the Common Fields', *Past and Present*, no. 33 (April 1966) are all republished together in R.H. Hilton (ed.), *Peasants, Knights and Heretics* (CUP, Cambridge, 1976).

4. An excellent collection of transcribed by-laws and court rolls relating to thirteenth- and fourteenth-century open-field practices are reproduced in W.O. Ault, *Open-Field Farming in Medieval England* (George Allen and Unwin, London, 1972).

5. M.M. Postan (ed.), *The Cambridge Economic History of Europe*: vol. 1 *The Agrarian Life of the Middle Ages*, 2nd edn (Cambridge University Press, Cambridge, 1966), p. 571.

6. The village by-laws were concerned with all aspects of agricultural management: the regulation of arable, pasture and meadow, harvesting and storage; fencing, boundaries, and roads and ways.

7. R.A. Butlin, 'Northumberland Field Systems', *Agric. Hist. Rev.,* vol. XII (1964), pp. 88, 99-120.

8. Thirsk, 'The Common Fields'. It can be demonstrated from documentary sources from 1150 onwards that communities often came together in order to adjust their field systems. It is not so clear that such adjustments involved total field realignment.

9. M. Chisholm, *Rural Settlement and Land Use* (Hutchinson, London, 1966), pp. 102–7.

10. J. Clapham, *A Concise Economic History of Britain from the Earliest Times to 1750* (CUP, Cambridge, 1949), p. 54.

11. Royal Commission on Historical Monuments, *An Inventory of the Historical Monuments in the County of Northampton, vol. II Archaeological Sites in Central Northamptonshire* (HMSO, London, 1979).

12. A.R.H. Baker, 'Field Systems of Southeast England' in A.R.H. Baker and R.A. Butlin (eds.), *Studies of Field Systems in the British Isles* (CUP, Cambridge, 1973).

13. M. Harvey, 'The Origins of Planned Field Systems in Holderness, Yorkshire' in T. Rowley (ed.), *The Origins of Open Field Agriculture* (Croom Helm, London, 1981).

14. H.S.A. Fox, 'Field Systems of East and South Devon Part 1: East Devon', *Transactions of the Devonshire Association*, vol. 104 (1972), pp. 81–135.

15. A.T. Gaydon (ed.), 'A History of Shropshire' in *Victoria County History*, vol. VIII (OUP, Oxford, 1968).

16. Beresford and St Joseph, *Medieval England*, pp. 45–8.

17. For a useful summary of this work see G. Jones, 'The Earliest Settlers in Britain' in A.R.H. Baker and J.B. Harley (eds.), *Man Made The Land* (David and Charles, Newton Abbot, 1973).

18. A.T. Gaydon, 'A History of Shropshire', p. 182.

19. C.C. Taylor, *Fields in the English Landscape* (Dent, London, 1974), pp. 96–7.

20. O. Rackham, *Trees and Woodland in the British Landscape* (Dent, London, 1976).

21. Taylor, *Fields in the English Landscape*, pp. 104–5.

22. R.A. Butlin, 'Field Systems of Northumberland and Durham' in Baker and Butlin, *Studies of Field Systems in the British Isles*, p. 112.

23. E. Nasse, *On the Agricultural Community of the Middle Ages and Inclosures of the sixteenth century in England*, translated by H.A. Ouvry, 2nd edn (Williams and Norgate, London, 1872).

24. F. Seebohm, *The English Village Community* (Longmans, Green and Co., London, 1883).

25. P. Vinogradoff, *Villainage in England* (OUP, Oxford, 1892).

26. F.W. Maitland, *Domesday Book and Beyond* (CUP, Cambridge, 1897).

27. H.L. Gray, *English Field Systems* (Harvard University Press, Cambridge, Mass., 1915).

28. C.S. Orwin and C.S. Orwin, *The Open Fields* (OUP, Oxford, 1938).

29. Beresford and St Joseph, *Medieval England*, p. 23.

30. J.G. Hurst (ed.), *Wharram, A Study of Settlement on the Yorkshire Wolds* (Society for Medieval Archaeology Monograph Series, no. 8, London, 1979), pp. 22–3.

31. Thirsk, 'The Common Fields'. All the quotations in this and the following two paragraphs are taken from the same paper.

32. Titow, 'Medieval England and the Open-Field System'.

33. H.S.A. Fox, 'Approaches to the adoption of the Midland System' in Rowley (ed.), *Origins of Open Field Agriculture*.

34. For a summary of this argument see J.C. Jackson, 'The Ridge-and-Furrow Controversy', *Amateur Historian*, vol. V (1961–2), pp. 41–53.

35. Baker and Butlin, *Studies of Field Systems in the British Isles*, pp. 34–5.

36. W.R. Mead, 'Ridge and Furrows in Buckinghamshire', *Geographical Journal*, vol. 120 (1954), pp. 34–42.

37. Taylor, *Fields in the English Landscape*.

38. D. Oschinsky, *Walter of Henley* (OUP, Oxford, 1971), p. 361. Walter of Henley emphasised that the furrows should be straight so that the surplus water could run down smoothly. It should be pointed out that there is some doubt about the accuracy of this reference as the copyists did not understand the technical content of this part of the manuscript.

39. P.A. Barker and J. Lawson, 'A Pre-Norman Field System at Hen Domen', *Medieval Archaeology*, vol. 15 (1971), pp. 58–72.

40. P. Rahtz and S. Hirst, *Bordesley Abbey: First Report on Excavations, 1969-73* (British Archaeological Reports, 23, Oxford, 1976), pp. 120–33.

41. J.G. Hurst (ed.), *Wharram*, p. 46.

42. P. Rahtz, 'Upton Gloucestershire, 1964–1968, Second Report', *Transactions of the Bristol and Gloucestershire Archaeological Society*, vol. 88 (1969).

43. Personal communication, S. Taylor and P. Rahtz

44. Beresford and St Joseph, *Medieval England*, p. 256.

45. C.C. Taylor, *Fieldwork in Medieval Archaeology* (Batsford, London, 1974).

46. G. Beresford, *The Medieval Clay-Land Village: Excavations at Gotho and Barton Blount* (Society for Medieval Archaeology Monograph Series, No. 6, London, 1975).

47. J.K.S. St Joseph, *The Uses of Air Photography*, 2nd edn (CUP, Cambridge, 1977), p. 158.

48. M.A. Aston and T. Rowley, *Landscape Archaeology* (David and Charles, Newton Abbot, 1974).

49. Taylor, *Fields in the English Landscape*.

50. Cambridge Committee for Aerial Photography.

51. J. Pickering, personal communication.

52. T. Rowley, *The Making of the English Landscape: Shropshire* (Hodder and Stoughton, London, 1972), p. 139.

53. H.C. Bowen, *Ancient Fields* (British Association for the Advancement of Science, London, 1961).

3 FORESTS, CHASES, PARKS AND WARRENS

Leonard Cantor

As we have seen, England during the Middle Ages was relatively well wooded. Virtually every manor possessed woodland, which was in common use by manorial tenants for collecting dead wood and grazing animals, and 'outwoods', woodland usually reserved for the landlord's use, were also quite common.[1] In addition, substantial areas of woodland were devoted to hunting, to satisfy the love of the chase of the Norman kings and barons and their successors. There were essentially four major hunting grounds, the *forest, chase, park* and *warren*. The *forest* was a large tract of country belonging to the Crown and subject to the forest law; the *chase* was, in effect, a private 'forest' which a few great nobles and ecclesiastical lords were allowed to create on their estates; the *park* was a securely enclosed area, relatively small in extent, and part of the demesne land of the lord of the manor; and the *warren* was essentially a game park filled with animals, principally hare and rabbits. In addition, the legal right of *free warren* was granted by the Crown to lords of the manor, entitling them to hunt the smaller game including the hare, rabbit, fox, pheasant and partridge over their own estates. Although the main purpose of these hunting grounds was the preservation of game for sport, they served incidentally to preserve substantial areas of woodland.

Forests

Although the Domesday Book does not give sufficiently specific information to calculate the extent of England covered with trees, we are left in no doubt about the wooded aspects of large tracts of England in 1086. Some clearing had taken place in the twenty years since the Conquest, for example in the four counties of Norfolk, Suffolk, Essex and Cambridgeshire, and doubtless the same was happening elsewhere, but this seems to have included areas where cutting was for timber rather than for assarting and, in any case, probably represented only a small proportion of the total woodland.[2]

Although woodland and royal forests were not identical, it is clear that many of the latter were for the most part well wooded. Although the Anglo-Saxon aristocracy had hunted extensively and had possessed game preserves, the practice was never institutionalised in the way the Normans created the royal forest, and indeed the Anglo-Saxons had by and large extended cultivation at the expense of woodland. In contrast, after the Conquest, the Normans introduced into England the Continental concept of the 'forest' as an area outside the common law of the land and under special laws and regulations designed to protect the king's hunting. In so far as these areas subject to the Norman forest law were well wooded, so incidentally woodland was also preserved.

However, the extent to which the legal forests were covered in woodland is open to some dispute. Most authorities, like Young and Birrell,[3] contend that the legal forests together accounted for a significant proportion of the total woodland cover of early medieval England. Rackham,[4] on the other hand, maintains that forests and woodland were not well correlated and that only about one-fifth of the legal forest was actually woodland; however, this seems an underestimate. It was certainly the case that some forests were almost entirely mountain and moorland, such as Exmoor, Dartmoor and the High Peak; others, such as the Wirral, contained relatively little woodland, and virtually all of them contained existing settlements, arable and pasture land, in which the owners had well-defined rights.

Clearly the Conqueror's love of hunting had by 1086 already led him to create other forests in addition to the New Forest and in 1087 the Anglo-Saxon chronicler wrote that King William made 'large forests for deer'.[5] As forests were outside common law, they are rarely specified or described in any detail. However, there are enough incidental references to make it certain that, by 1086, a substantial number of forests had come into being including those in the counties of Berkshire, Buckinghamshire, Cheshire, Dorset (Plate 2), Essex, Gloucestershire, Hampshire, Herefordshire, Huntingdonshire, Lancashire, Northamptonshire, Oxfordshire, Shropshire, Staffordshire, Surrey, Sussex, Warwickshire, Wiltshire and Worcestershire.[6]

The main purpose of the forest land was to protect wild animals for the king's hunting, namely the fallow deer, the red deer, the roe deer (until a judicial decision in 1339 removed it from the list because it was supposed to drive away other deer), and the wild boar. However, the beast of the chase *par excellence* was the fallow deer. In order to render the forest law effective over such considerable areas, a large administrative machinery, including means of enforcement by courts and justices, developed alongside but apart from the common

Plate 2: The Isle of Purbeck in 1585: this area, which was a royal forest by the time of Domesday, still contained deer five centuries later

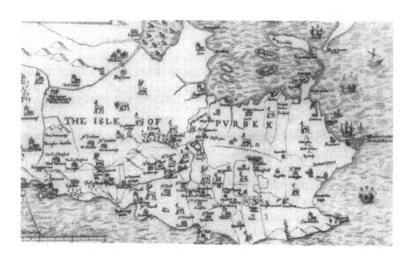

law, probably soon after the Conquest and certainly by the beginning of the twelfth century.[7] Inevitably, the administration of the forest law, particularly in the early Middle Ages when it was at its harshest, was liable to abuse by local forest officials and aroused bitter resentment among the king's subjects. As we shall see, for 150 years after the Conquest, the forest laws might fall into disuse at times of civil discord, whereas at times of stronger central government they would be stringently applied. The application of the forest laws is of interest to us only in so far as it indicates those areas of the country which were, at various times, under its jurisdiction. However, there is no doubt that the forceful application of the forest laws prevented landowners and farmers in areas covered by it from agricultural improvements that involved clearing the land and extending the cultivated area. Indeed, tillage of the areas already under cultivation must have been very difficult as the cultivator was not allowed to erect hedges to prevent the deer from gaining access to his crops. Indeed, if the farmer found deer eating his corn and drove them off with his dogs, he was liable, in the earlier part of the Middle Ages at least, to summary conviction at the forest courts.

More important, however, was the fact that the full application of the forest law inhibited agricultural extension and improvement over considerable areas of the country, a situation very different from the Continent and especially France where the royal hunting preserves were much more modest in extent. This inhibition on economic development, especially at the time of population growth, inevitably built up a great deal of pressure upon the Crown to which, from the late twelfth century onwards, it was forced progressively to bow and to contract the areas within which the forest law applied. On the other hand, it could be argued that the forest law was to some extent a deterrent to widespread disafforestation which might otherwise have resulted in the spread of arable farming into areas unsuited to it.[8]

In addition to those parts of the forest where landowners and farmers already held traditional rights of cultivation and stock rearing, other agricultural activities developed as a result of specific territorial grants and franchises made by the Crown to a favoured few of its subjects. The importance of these grants can be gauged from the probability that in several parts of Lancashire, for example, much of the development of towns and villages in the period between 1066 and the early thirteenth century was directly associated with them,[9] and the same is doubtless true of other forest areas. These grants were generally made to three major groups of beneficiaries: to monastic houses, to nobles and courtiers in the royal favour, and to forest officials. The religious houses were intimately associated with the forests and many monasteries possessed specific privileges conferred upon them by charter: for example, the abbeys of Roche in Yorkshire, Welbeck in Nottinghamshire, Merivale in Warwickshire, and St Mary, Leicester – among others – held rights in the Peak Forest, as had Chertsey Abbey, Surrey in Windsor Forest, and Newstead Priory, Nottinghamshire, in Sherwood Forest.[10] These rights generally allowed them to fell timber for building purposes, to collect dead wood and undergrowth for fuel, and to graze their cattle and to turn their pigs into the forest to forage. Occasionally the abbey would be given gifts of venison, in the form of the tithe of a hunt. The grant of sections of the forests to powerful clerics or nobles for use as private chases was, as we shall see, a not infrequent occurrence.

The royal hierarchy of forest officials included two justices of the forest – one for those north of the Trent and one for those to the south – wardens in charge of individual forests, and subordinate foresters, usually local landowners who served as foresters in fee over particular parts of the forest. The wardens often had very considerable holdings within the forests, as in the cases of the Warden of the Forest of Dean in Gloucestershire and the Warden of Savenake Forest

Figure 3.1: The royal forests in the thirteenth century

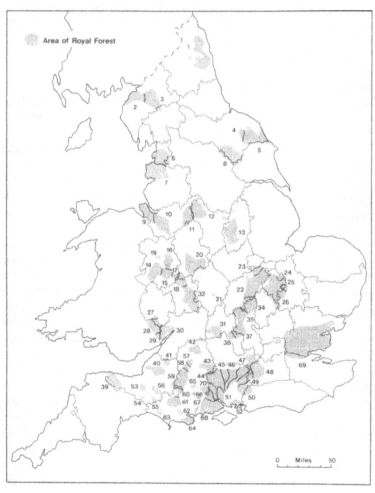

Notes

50 Aliceholt and Wolmer	70 Buckholt
2 Allerdale	20 Cannock
7 Amounderness	29 Chepstow
49 Bagshot	57 Chippenham
62 Bere	44 Chute
51 Bere Ashley	68 Clarendon
52 Bere Porchester	30 Dean
37 Barnwood	10 Delamere
61 Blackmore	69 Essex
42 Braden	47 Eversley

39 Exmoor
4 Farndale
32 Feckenham
45 Freemantle
8 Galtres
60 Gillingham
65 Groveley
19 Haughmond
27 Haywood
26 Huntingdon
3 Inglewood
28 Irchenfield
21 Kenilworth Park
41 Kingswood
18 Kinver
14 Longforest
6 Lonsdale
11 Macclesfield
58 Melksham
40 Mendip
67 Milcet
17 Morfe
54 Neroche
55 To Neroche
68 New Forest

53 North Petherton
1 Northumberland
46 Pamber
12 Peak
5 Pickering
63 Powerstock
64 Purbeck
24 Ramsey
22 Rockingham
23 Rutland
34 Salcey
43 Savernake
59 Selwood
13 Sherwood
15 Shirlet
38 Shotover
25 Somersham
56 Somerton
35 Whittlewood
48 Windsor
50 Wolmer and Aliceholt
16 Wrekin
31 Wychwood
9 Wirral

Source: Based on Margaret Bazeley, 'The Extent of the English Forest in the Thirteenth Century', *Transactions of the Royal Historical Society*, 4th ser., vol. 4 (1921), p. 165.

in Wiltshire, an office that for most of the Middle Ages was in the hands of the Esturmy family.[11]

During the reign of Henry I (1100-35), the royal forests were somewhat extended, certainly in Yorkshire and possibly in Cumberland also.[12] Inevitably, during the chaotic years of Stephen's reign which followed (1135-54), the administration of the forests could not be efficiently maintained and it is likely that the extent of the forest was smaller than during the reign of his predecessor.[13] On his accession to the throne, Henry II (1154-89), an efficient and ruthless man, restored the effective administration of the forest laws and gradually extended the royal forests to the greatest area they were ever to attain. Thereafter, the remainder of the Middle Ages was to witness their gradual but inevitable decline. Nevertheless, by the thirteenth century they still occupied about one-fifth of the country (Figure 3.1), being concentrated particularly in the south central counties, the east and west Midlands, the Peak District, and parts of Lancashire, Yorkshire, Cumberland and Northumberland. By contrast, parts of the country were wholly devoid of forest, notably the south-east, the south-west, East Anglia and Lancashire.

The trees that made up the natural woodland cover of the medieval forest were principally the oak, birch and alder. Oak was to be found virtually throughout the country and was the commonest and most valuable timber tree; for this reason, it is the tree most frequently mentioned in contemporary documents. The oak, birch and alder were valued because their bark was stripped for tanning. Other fairly common species included the ash, field maple and lime; less common were the beech, which was native only on the southern half of England, and the elm and the pine, which did not become widespread before the eighteenth century.[14] The woodland cover of the medieval forest must have presented a relatively impenetrable aspect as rides seem to have been absent or uncommon until the post-medieval period. However, there were probably impermanent tracks wending their way through the forest.[15]

In places, the forest woodland was carefully and intensively managed. Here, as in private woodland, a cropping rotation was used to ensure that the felled timber was replaced. This mostly took the form of 'coppice with standards', that is standard trees and underwood. The former, consisting principally of oak, ash, hazel, maple, lime and crab-apple, yielded timber at regular, if often quite lengthy, intervals which was used principally for the construction of buildings and agricultural implements. The latter, consisting of whole or split sallow, or hazel rods or laths left from oak timber, were tied together with young sallow shoots or string, and used for wattle and daub and for making hurdles and fences.[16]

Apart from providing the king with hunting and with wood, the royal forests were also an important source of other revenues. Payments of rent by landowners in the forests brought in considerable sums of money, as did grants of pannage for pigs and herbage for domestic animals. Some forests contained vaccaries, or cattle farms, for which rents were paid; and the sale of hides, the cutting of turf and the making of salt were all to be found within the forests, which also provided a copious source of venison, eaten in very large quantities by the king and the court. There were also payments by landowners in the forests for such special purposes as setting up sheep folds, allowing domestic animals to run free in the forest at night throughout the year, and for mining iron. In addition, forest courts brought in revenue in the forms of fines and confiscations.[17] However, the forests were also expensive to maintain as officials like foresters and verderers had to be employed and it was often necessary to provide food for the deer in winter. Although the extant records of the period do not allow an accurate estimate to the made of the annual income which the Crown derived from its forests, there is little doubt that by the mid-thirteenth century it was very considerable.[18]

The royal forests were also of importance in supplying directly a variety of products used by the Crown. Clearly of prime importance was timber, used for repairing royal castles and houses, fitting ships for the navy and making weapons for the king's armies, as well as supplying wood for charcoal-making and as a source of fuel for heating.[19] As we have seen, the deer, apart from being the main quarry of the royal huntsmen, were also a most valuable source of meat for the royal table, for which purpose they were killed in very large quantities. They were also used to stock royal parks or granted as gifts by the king to favoured nobles to stock their parks. In addition, horses were raised in some of the forests, including the New Forest which had a number of stud farms. Finally, royal forges for the working of iron ore in the Forest of Dean and lead-smelting in the forest of High Peak were typical of the industrial activities of some of the forests. In short, during the thirteenth century, far from being separated from the expanding economic activities of England, the royal forests were an integral part of the agrarian and industrial life of the period.[20]

A good example of the way in which forests were exploited in their heyday, the early Middle Ages, is provided by the four medieval forests of Cheshire.[21] These consisted of Wirral Forest in the north-west of the county; Delamere and Mondrem, which were contiguous but administered separately, in central Cheshire; and Macclesfield Forest in the east of the county (Figure 3.2). Together they made up a substantial part of Cheshire, each being between 45 and 85 square miles in extent.

The four forests varied considerably, both regarding the distribution of woodland cover and also the level of economic activity. Wirral Forest, for example, was unlike most medieval forests in having a relatively high level of population and agricultural development in 1086 and little recorded woodland cover. Delamere and Mondrem, on the other hand, had relatively few settlements, located mainly on the flanks where the rivers Gowy and Weaver cut their valleys; these river valleys were also wooded, while, by contrast, the central Triassic sandstone uplands which made up much of the area contained few vills and were given over mainly to heath. Macclesfield Forest, too, had a central area of exposed moorland, devoted to heath and poor pasture; only the adjacent slopes were wooded and contained the few

Figure 3.2: The medieval forests of Cheshire

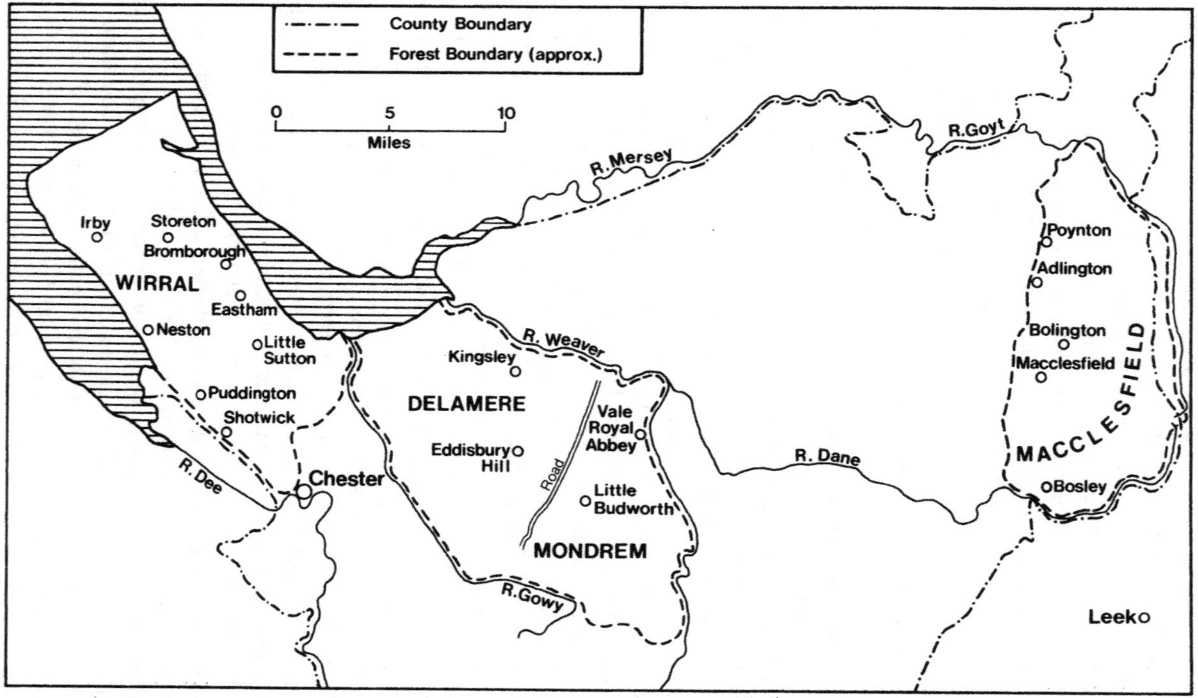

Source: Based on *Victoria County History, Cheshire*, vol. II (1979), p. 168, by permission of the general editor.

Domesday settlements. All the forests seem to have been plentifully stocked with both fallow and red deer; in 1363, for example, the Black Prince ordered one hundred harts and one hundred bucks to be taken from the Wirral Forest, salted and packed in barrels and sent to him at Bordeaux.

An important activity in the Cheshire forests was the keeping of cows, in the form of royal vaccaries, which were located mainly in the upland areas. These are mentioned in Macclesfield Forest in 1285 and 1286, but by the middle of the fifteenth century, in common with other royal forests' pastures, these were being leased for rent to local landowners. In 1442, for example, Sir Thomas Stanley leased pastures in Macclesfield Forest, which remain in the hands of the Stanley family for over two hundred years. Other uses to which the forests were put included pannage, the feeding of pigs on acorn mast; the tanning of bark, principally from the oak, birch and alder; and the sale of turbuary, or peat. Occasional royal parks were also to be found in the forests, as for example Shotwick Park in Wirral Forest, and Macclesfield, both of which were in existence by the middle of the thirteenth century.

In Cheshire, as throughout the country, the forests gradually declined through the later Middle Ages and the woodland cover was steadily eroded, principally by assarting. References to such encroachments on the forests are found as early as the twelfth century when certain barons, knights and free tenants had the right to assart within the limits of their husbandry in the forests. By this time, the Crown seemed to have accepted the inevitable and was charging a fee for assarting in the forests of Delamere, Mondrem and Macclesfield. The same was true of Wirral Forest where, in 1275, landowners were being charged an annual rent of five shillings for an acre of heath and six shillings for an acre of woodland. In this and other ways, the Cheshire forests gradually declined in extent and effectiveness. Nevertheless, for centuries during the Middle Ages they remained under the forest laws and contained 'wilderness' areas which were the haunt of fugitives and outlaws. In the later fourteenth century, for example, the Wirral Forest was the resort of bands of armed men.

Wirral Forest was the first of the four Cheshire forests to be disafforested in 1376, and as it was relatively well populated and much given over to agriculture, its early disafforestation was probably inevitable. The other three forests remained technically subject to the forest law throughout the Middle Ages and it was not until the early seventeenth century that Mondrem became effectively excluded from the

forest area. At the time of the Civil War, both Delamere and Macclesfield Forests contained deer, but those in Delamere were destroyed
then and never replaced. Macclesfield Forest effectively came to an
end in the 1680s when the Earl of Derby acquired outright pastures
in the forest which had been leased to his family since the fifteenth
century; thereafter, the forest survived only as an administrative area
in connection with the court of the manor and forest. The last forest
to be officially disafforested was Delamere which continued its legal
existence until as late as 1812.

In the country at large, as in Cheshire, the later Middle Ages witnessed the slow but sure decline of the forests. Their legitimiate boundaries became a matter of dispute between the Crown and the barons
at the end of the thirteenth century and the king was forced to agree
to new perambulations. These took place in and about 1300, and are
illuminating for the light they cast on the extent of the forest at that
time and the claims for disafforestation of the people within their
bounds. The Forest of Cannock in Staffordshire, for example, had
in 1286 encompassed virtually all the land from the river Trent south
to Wolverhampton and Wednesbury (Figure 3.3). However, in the
perambulation of 1300, large numbers of vills south of Watling Street
claimed disafforestation on the grounds that much of the forest had
been unjustly created by Henry I. If conceded, these claims would
have reduced the forest to seven small areas, or 'hays'. In practice, they
were judged to be false and it was not until 1327 that they were
accepted and the forest much reduced in area.[22]

By about 1334, the area of the royal forests in the country as a
whole had shrunk to about two-thirds of what it had been in 1250
(Figure 3.4). During this period, most of the individual forests had
diminished in extent and some, like the Forest of Northumberland
and Allerdale Forest in Cumberland, had disappeared completely.
The largest forests still in existence were the New Forest, the Forest
of Dean, Sherwood Forest, Windsor Forest, Inglewood in Cumberland and Pickering in Yorkshire. The distribution of the royal forests
at this time was very uneven and was very similar to the pattern that
existed a century earlier.

It is impossible precisely to delineate, in geographical terms, the
continued decline of the royal forests in the later Middle Ages, as it
was not so much that there was a substantial decrease in the actual
area to which the forest law was nominally applied, but rather that
its application became less and less effective so that increasing areas
were 'forests' in name only. This change in the status of the forests

Figure 3.3: Cannock Forest, 1286-1300

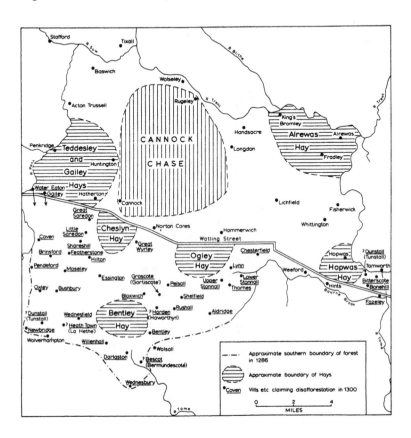

Source: Based on L.M. Cantor, 'The Medieval Forests and Chases of Staffordshire', *N. Staffs. Journal of Field Studies*, vol. 8 (1968), p. 45, by permission of
the editor.

was that they no longer commanded the serious attention of the
Crown, sufficient to ensure the enforcement of the forest law, because
the contribution it made to the royal revenues had become less and less
significant. During the fourteenth century, a system of public finances
had developed, based upon taxes levied on the growing wealth of the
nation, which eventually completely eclipsed revenues obtained from
the royal forest, which in any case had been reduced by disafforestation. However, the basic administration structure of the forests
continued throughout the Middle Ages, headed by the two justices for

Figure 3.4: The royal forests, 1327–1336

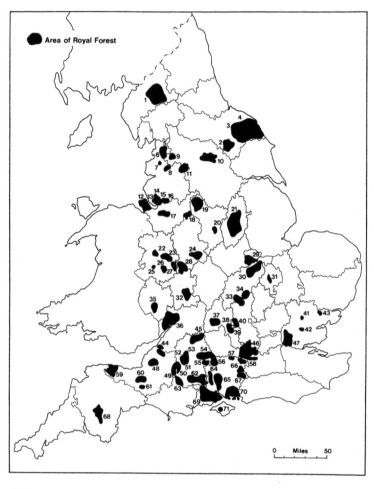

Notes

66	Alice Holt	24	Cannock
58	Bagshot	53	Chippenham
70	Bere by Porchester	56	Chute, Hampshire
65	Bere by Winchester	55	Chute, Wiltshire
40	Bernwood	62	Clarendon
11	Blackburn	16	Croxteth
6	Bleasdale	68	Dartmoor
9	Bowland	36	Dean
45	Braden	17	Delamere
64	Buckholt	20	Duffield

47 Essex
59 Exmoor
32 Feckenham
 8 Fulwood
 2 Galtres
63 Gillingham
41 Hatfield
35 Hereford Hay
19 High Peak
 1 Inglewood
71 Isle of Wight
43 Kingswood, Essex
44 Kingswood, Somerset
28 Kinver
10 Knaresborough
22 Lithewood
18 Macclesfield
51 Melksham
48 Mendip
27 Morfe
 7 Myerscough
61 Neroche
69 New Forest
60 North Petherton
57 Pamber
52 Pewsham

 4 Pickering
 5 Quernmore
30 Rockingham
29 Rutland (or Leighfield)
34 Salcey
54 Savernake
49 Selwood, Somerset
50 Selwood, Wiltshire
21 Sherwood
26 Shirlet
38 Shotover
14 Simonswood
 3 Spaunton
39 Stowood
25 Stretton
15 Toxteth
31 Wauberghe
23 Wellington
13 West Derby
33 Whittlewood
46 Windsor
12 Wirral
67 Woolmer
42 Writtle
37 Wychwood

Source: Based on Nellie Neilson, 'The Forests' in James F. Willard and W.A. Morris (eds.), *The English Government at Work, 1327-1336* (Medieval Academy of America, Cambridge, Mass., 1940), vol. 1, map V.

the forests north and south of the Trent and the wardens for individual forests. Moreover, the main uses to which the forests had always been put — hunting and a variety of economic returns — continued as before, albeit on a smaller scale: farms and wastes were rented and timber and underwood were sold. Hunting within the forests was still restricted to the king and to those to whom he gave licence, though the enforcement of this restriction became more and more difficult, especially after the plagues of the mid-fourteenth century.

By the late fifteenth century, the forests, so important a legal and geographical institution in the early Middle Ages, had become a shadow of their former selves. Although there was a revival of royal interest in them by Charles I in the second quarter of the seventeenth century as a means of raising money,[23] it was short-lived and the Civil War and its aftermath resulted in increasing destruction of woodland cover. Subsequent events have conspired to reduce the once-extensive forests to residual remnants like Epping Forest and the New Forest, which are fortunately still a feature of contemporary landscape. In retrospect, the chief result of the introduction of the concept of the forest by the Normans and the application by them and their successors of the forest law was to delay the inevitable clearing of the woodland cover which was so prominent a feature of the English landscape at the beginning of the Middle Ages.

Chases

As we have seen, the chase was, in effect, a private forest. It was an area of land, often quite extensive, over which local magnates, usually nobles or great ecclesiastics, were given rights of hunting by the king. No one might hunt in a chase without permission of the holder of the franchise and, as in the case of the forests, not only the deer but also the wolf and the boar were reserved.[24] The chase was normally subject to common law and not forest law, and the rights of the chase seem to have varied somewhat: in some cases the owners enjoyed only limited rights of protecting the deer and venison,[25] while more often the holders of the franchise imposed restrictions on people living within the chase, enforced by their own officials, that were little different from those which applied to the royal forests. During the Middle Ages, and indeed since, the terms 'chase' and 'forest' have been sometimes used synonymously. However, the legal difference between them — the former belonged to the Crown and was subject to forest law and the latter to a private individual and was subject to common law — was normally quite distinct. Occasionally, however, even this distinction became somewhat blurred. For example, from 1285 onwards the Earls of Lancaster, in respect of their chase of Leicester, were empowered by the king to enforce the whole body of the forest law as found in operation over the royal forests and were entitled to appoint their own justices and to hold forest courts.[26] A similar situation obtained in the other Lancastrian chases of Bowland and Blackburn in Lancashire and Needwood in Staffordshire.[27] This state of affairs continued until 1399 when a Lancastrian king came to the throne, in the shape of Henry IV, the Duchy of Lancaster became Crown land, and the Lancastrian chases became truly royal forests.

At various times in the Middle Ages, there were at least twenty-six chases in existence, including Blackburn and Bowland in Lancashire; Cannock, Needwood and Pensnett in Staffordshire; Copeland, Nichol and Geltsdale in Cumberland; Cranborne in Dorset; Dartmoor in Devon; Duffield Frith in Derbyshire; Hatfield in Yorkshire; Enfield

Figure 3.5: The medieval chases of England

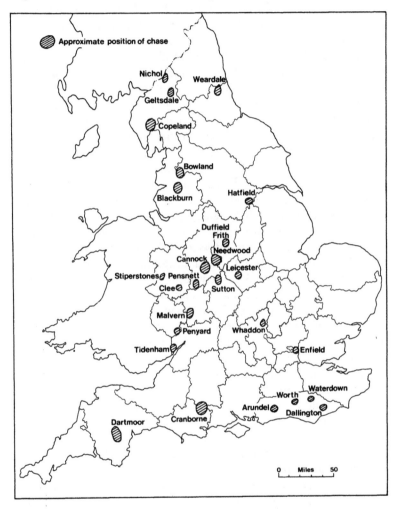

in Middlesex; Leicester; Penyard in Herefordshire; Clee and Stiperstones in Shropshire; Sutton in Warwickshire; Tidenham in Gloucestershire; Weardale in Durham; Whaddon in Buckinghamshire; and Arundel, Worth, Waterdown and Dallington in Sussex (Figure 3.5).

The Earls of Lancaster in the fourteenth century owned five chases, at Blackburn, Bowland, Duffield, Leicester and Needwood, and they exerted a degree of supervision and control over these extensive tracts of country which was at least as complete as that exercised by the Crown over royal forests. As in the royal forests, the owners of the chase appointed their own officials: in the case of Leicester Chase, for example, they consisted of a master forester, a forest receiver who was responsible for the financial administration of the chase, three foresters and various other officials.[28] Like the forests, the chase brought in considerable revenues to their owners: in the case of Needwood, for example, which was largely situated on heavy, fertile Keuper Marl soil giving rise to fine grassland and excellent timber, both the sale of wood and the lease of agistment were profitable activities.[29] Once the Lancastrian chases came into royal hands with the accession to the throne of Henry IV in 1399, they seem to have suffered mixed fortunes. Leicester forest, as it had become, went into decline largely due to royal neglect; in the case of Needwood, however, the administrative machinery set up by the Duchy continued to function well into the fifteenth century and as late as 1559 it was still 23½ miles in compass and contained 7,869½ acres. It was not finally disafforested until the beginning of the nineteenth century.[30]

Other chases were, for various periods, in the hands of other landed magnates, both lay and ecclesiastic. Thus, Tidenham Chase in Gloucestershire, which was at one time in the Forest of Dean, had by the early thirteenth century come into the hands of the Earl of Pembroke;[31] and Sutton Chase in Warwickshire was taken from Cannock Forest in 1125 and granted by Henry I to the Earl of Warwick.[32]

Cannock Chase, like Sutton, was also taken out of a royal forest and this seems to have a common way of creating chases. Cannock belonged to the Bishop of Coventry and Lichfield who was given permission to turn his manors of Cannock and Rugeley into a private forest in 1290.[33] At this time, it enclosed an area of about forty square miles and, although chases were not normally enclosed by earthworks, part of the western edge of Cannock Chase was delimited by a boundary bank, at Huntington a few miles north of Cannock. It seems likely that such banks were erected only when no other obvious topographical feature was available. The chase remained in the bishop's hands throughout the Middle Ages and was taken over by the Crown at the Dissolution.

Finally, the history of Cranborne Chase is a more complex one, in that it passed in and out of royal hands, including those of King John who hunted there frequently, but remained with the de Clare Earls of Gloucester for much of the Middle Ages. It was given by William Rufus to Robert, Earl of Gloucester at the end of the eleventh century, at which time it probably encompassed several hundred square miles extending from Dorset into Wiltshire. Consisting largely of a chalk plateau, Cranborne Chase was never very heavily wooded and consisted both of open rolling downland and also numerous woods and copses scattered with vills surrounded by arable fields. Some idea of the extent of the medieval woodlands on the chase is given by an inquisition postmortem on Gilbert de Clare in 1296, which indicates that there were 1,433 acres of woodland where the pasture was worth 100 shillings a year which could not be sold because of the deer.[34] As with all the other chases, Cranborne gradually shrank in size but continued to be used as a hunting ground until as recently as 1828. A vivid picture of its Victorian appearance is conjured up in *Tess of the d'Ubervilles* by Thomas Hardy, who described it as 'a truly venerable tract of forest land, one of the few remaining woodlands in England of undoubted primeval date wherein Druidical mistletoe was still found on aged oaks, and where enormous yew-trees, not planted by the hand of man, grew as they had grown when they were pollarded for bows'.

Parks

As we have seen, the principal features that distinguished the park from the other medieval hunting grounds, the forest and the chase, were its size and the fact that it was securely enclosed. A very common feature of the medieval landscape, it was normally fairly small, being usually between 100 and 200 acres in size, though some parks were very much larger: for example, the royal parks of Woodstock in Oxfordshire, and Clarendon in Wiltshire, were both seven miles in circuit.[35] The enclosure to retain the deer normally consisted of a combination of a substantial earth bank, topped by a fence of cleft oak stakes, with an inside ditch which together consistuted a formidable barrier (Plate 3). In some areas, where free stone was easily available or the landowner could afford to transport stone, the wooden fence might be replaced by a stone wall, as in the case of Woodstock

Plate 3: The substantial remains of two medieval Dorset park banks: *a* Frome St Quintin, north-west of Dorchester (above) and *b* Ryehill, Wimborne St Giles (below)

and Moulton Park, Northamptonshire.[36] Occasionally a quickset hedge would serve in place of a fence and where the topography was suitable, for example just below the crest of a steep slope, the paling fence alone might suffice. Water seems to have constituted an effective barrier to deer and some parks were partly circumscribed by rivers or marshy areas, as in the case of Sonning and Hamstead Marshall Parks in Berkshire where the northern boundaries were formed by the Thames and Kennet respectively. The park perimeter usually followed a compact course to keep its length down to a minimum and a roughly elliptical or circular shape was common. The circuit was broken by gates for passage in and out of the park and occasionally by 'a deer leap'. The latter was a device which enabled deer to enter the park but not to leave it and, for this reason, was a privilege eagerly sought after but reluctantly granted by the Crown in whose possession wild deer were vested.

The park was owned by the lord of the manor and was part of his demesne lands, Typically, it consisted of 'unimproved land', beyond the open fields and on the edge of the manor, and included woodland to provide covert for the deer. The main functions of the park were to provide hunting for the lord of the manor and a source of meat, though, as we shall see, it was put to a variety of other uses. The medieval park was therefore quite different in appearance from the later 'amenity' parks which were landscaped in order to improve the surroundings of the great houses of the eighteenth century.

Initial emparkment was usually on a small scale, since the construction of the park pale was a major operation involving a great deal of labour, which few landowners could command on a large scale. However, although most parks started out as relatively small areas of demesne woodland, many of them were extended piece by piece over the course of several centuries, depending on the fortunes of the owners. Woodland was always an essential part of the park for as John Manwood, the Elizabethan authority on hunting and the forest wrote, 'it must be stored with great woods or coverts for the secret abode of wild beasts and also with fruitful pastures for their communal feed'.[37] It does not seem that it was necessary for a landowner to obtain a licence to create a park unless it was close to or within a royal forest, in which case it might interefere with royal forest rights.[38] The principal beast of the park was the fallow deer, though red deer were also quite common and many favoured park owners, when first creating their parks, received gifts of deer from the Crown with which to do so. In Buckinghamshire, for example, Richard Montfitchet was given 100

live does and bucks from Windsor Forest in 1202 for his park at Lang-
ley Marish, and in 1222 Robert Manduit was granted five stags from
Salcey Forest to stock his park of Hanslope.[39]

Like the forest, the medieval park was essentially a creation of the
Norman kings and barons and it, too, testified to their love of hunting.
There were enclosures or 'deer-folds' in existence in the Anglo-Saxon
period, for example at Ongar in Essex, but there were certainly many
fewer in number than the parks were to become and may not have been
so securely enclosed. In 1086, there were at least 36 parks in existence,
35 recorded in the Domesday Book and another at Bramber in Sussex,
belonging to William de Braiose.[40] These parks were held by the king,
the Bishops of Bayeux and Winchester, other powerful clerics, and
William's personal supporters. The greater number was in the south-
eastern part of the country and around the home counties; none is
mentioned in the Midland counties, except for three in Worcestershire,
and none at all in the north.[41] However, the Domesday total is certainly incomplete and others must have been created between the
Conquest and 1086. The Crown and the great magnates, lay and ecclesiastical, continued to be the owners of the largest numbers of parks
throughout the Middle Ages. The Crown was by far the largest single
proprietor and its most important parks tended to be associated with
the royal forests like Woodstock in Wychwood Forest, Gillingham in
Dorset, and Clarendon in Wiltshire. They were much frequented by
monarchs, especially in the early Middle Ages, and frequently included
a hunting lodge, as in the case of the so-called 'King's Court Palace' in
Gillingham Park, Dorset.[42] The great landowners with many large
estates were also prolific creators of parks which were generally situated
in their principal manors. For example, at various times in the Middle
Ages, the Earls of Lancaster held 45 parks, principally in Lancashire,
Staffordshire and Leicester, the Dukes of Cornwall 29, the Earls of
Arundel 21 and the Earls of Norfolk 15. The wealthy bishops were also
owners of numerous parks, the Bishop of Winchester possessing 23 in
his estates, the Archbishop of Canterbury 21 and the Bishop of Durham
20. The great religious houses also possessed many parks, notably the
Abbeys of Glastonbury, Bury St Edmunds and Peterborough.[43]

In the century and a half after the Conquest, a growing number of
parks was created, including Woodstock in 1113, and Devizes, but the
relative lack of economic development during this period and the power
of the Crown in enforcing the forest law over large parts of the countryside inevitably limited their development. It was the century and a half
between 1200 and 1350 that witnessed the great expansion in the

number of parks. A period of agricultural development and growing population, it produced sufficient surplus wealth to enable many noble and knightly families to indulge in their love of hunting by creating parks on their estates. They may also have come to be regarded, like fortified houses, as status symbols, manifestations of conspicuous consumption during a period which could clearly afford them.[44] Their owners could call upon the feudal services of their tenants to maintain and repair the park bank and pale and despite growing land hunger there was still wooded, less fertile land available in many manors which was well suited to imparkment. More generally, increased demands for disafforestation during this period resulted in the gradual concessions made by the Crown and land becoming available, much of it wooded and ideally suited to imparkment. As we shall see, there appears to be a close correlation between the distribution of parks and areas of woodland.

As in the case of several of the major features of the medieval landscape, the plagues of the middle of the fourteenth century marked a watershed in the fortunes of the park. Particularly after the Black Death, a slow decline set in from which it never recovered. Many deer parks fell out of use, the herds dwindled and labour was no longer available to maintain them properly. As a consequence, many were disparked or existed in name only and, increasingly, pasture within them was leased out for long periods, a trend which followed the decline of direct demesne farming. On the other hand, in various parts of the country tracts of arable and pasture land which could no longer be properly farmed because of shortage of labour were converted into parks. However, these parks were generally much larger than their predecessors of the early Middle Ages and were probably neither managed as intensively nor as securely enclosed. Indeed, many of them must have been conceived from the beginning as amenity parks rather than hunting parks. In addition, some small, early medieval parks greatly extended their boundaries or, having fallen into disuse, were reimparked during this period.

As an integral part of the demesne land of the manor, the medieval park during its heyday was as carefully and economically managed as any other part of the manor. Indeed, in many cases its economic contribution may well have outweighed its sporting value. Thus, the land within the park pale was put to many other activities, as in the case of Madeley Great Park, Staffordshire.[45] This park, which was certainly in existence in 1272 and may well have originated after about 1204 when north-west Staffordshire was probably disafforested, was extremely long-lived and continued well into the modern period, being still in existence at the end of the seventeenth century. In addition to hunting, Madeley Great Park provided three other main sources of revenues for its owners, the Earls of Stafford — agistment (pasturage), wood and turbary (peat) — none of which activities would interfere with the deer. In addition, stone-mining, pannage, rabbits and fishing were also more occasional sources of revenue. To balance the revenues derived from the park, expenditure included the wages of the parker, £3 0s 8d in the year 1439/40, for example, and the cost of repairing the park pale. Other uses to which parks were put included stud farms, as in the royal parks at Princes Risborough in Buckinghamshire and Haywra Park, Knaresborough in the West Riding of Yorkshire, and parts of the park might occasionally be ploughed up for tillage, as in the case of the Earl of Cornwall's Cornish parks in the late thirteenth century and in Pulham Park, Norfolk in 1251.[46]

Finally, fish-ponds were frequently situated in parks to provide additional protein in the form of fish and, presumably, to help drain the land. Indeed, the *fish-pond* was a not uncommon feature of the medieval landscape, both within and without parks. It was usually made by building an earth bank across the line of a stream and, where the valley sides were not steep enough, two additional embankments might be built parallel to the stream.[47] Although most sites consisted of a single pond, others comprised groups of three or four, one leading to another, somtimes relatively large in size. They took a variety of forms, determined in large part by local topography: in central Northamptonshire, for example, seven main types have been identified,[48] and a similar diversity has been found in Warwickshire and Worcestershire.[49] Although many medieval fish-ponds have long since lost their water, traces of banks and dams can often still be found in the modern landscape.

Parks were to be found throughout the country in medieval England, though, as in the case of moated homesteads, they were more densely located in some counties than in others. As can be seen in Figure 3.6, they were thickest on the ground in the west Midlands, in Staffordshire and Worcestershire; in the Home Counties, in Essex, Hertfordshire, Buckinghamshire and Surrey; and in Sussex. The counties with the fewest parks were in the remote parts of the country, in Northumberland, Cumberland, Durham, Devon and Cornwall; and in East Anglia, in Norfolk, Cambridgeshire and Lincolnshire. The comparison with the distribution of moated homesteads is striking and in both cases there is a marked correlation between areas of high woodland cover, as

Figure 3.6: The density of medieval parks

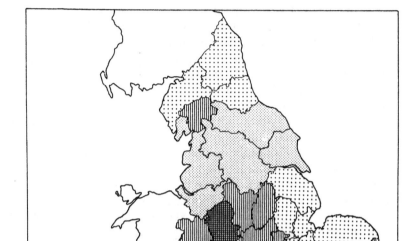

1 park per less than 10,000 acres	
1 park per 10-20,000 acres	
1 park per 20-30,000 acres	
1 park per 30,000 acres or more	

0   Miles   60
0   Kms.   100

Source: Based on L.M. Cantor and J. Hatherly, 'The Medieval Parks of England', *Geography*, vol. 64 (1979), p. 75, by permission of the editor.

evidenced by the Domesday Book, and areas containing many parks. Thus, areas of relatively fertile soils, already settled and cultivated at the time of Domesday, would be unsuitable for hunting without considerable disruption of agricultural activities and changes in the landscape, such as the planting of trees, and would therefore be generally avoided by imparkers. Areas already wooded, on the other hand, especially those recently freed of the restrictions imposed by the forest law, would be eminently suitable for the creation of parks. It is perhaps mainly for this reason that the two counties with the largest number of parks, Essex and Staffordshire, containing at least 98 and 88 respectively, were well wooded and, at the beginning of the Middle Ages at least, contained very large tracts of land subject to forest law which were subsequently disafforested. Other counties with substantial numbers of parks were the West Riding of Yorkshire with at least 73, Somerset with 71, the North Riding of Yorkshire with 67, Hampshire with 63, and Gloucester with 60.

Within individual counties, a number of factors helped to determine the distribution of parks, as is shown in the case of Staffordshire and Warwickshire. In the former, much of the county was relatively poor, undeveloped and unsuitable for arable farming; as a consequence, parks were numerous and well distributed. However, they are markedly absent in some parts of the county: along the cultivated tracts of the middle Trent valley, between Stoke-on-Trent and Stafford; in the royal forests of Cannock, between Stafford and Cannock, and Kinver in the extreme south-west of the county; and in the bleak moorlands of the north. Although royal forests did contain a few royal parks, by and large they inhibited their development, as can also be seen in the case of Dorset, where the forests of Blackmore, Gillingham and Purbeck contained very few parks.[50] On the other hand, the private parks or chases, especially those of the Earls of Lancaster, frequently possessed numerous parks, as in the case of Staffordshire where Needwood Chase, between Burton upon Trent and Stafford, contained eleven Lancastrian parks in the fourteenth century. In the county of Warwickshire, the influence of woodland cover on the distribution of medieval parks is strikingly demonstrated. The county fell, naturally, into two sections in the Middle Ages: that north of the river Avon, known as 'Arden Warwickshire', relatively poor, infertile and well-wooded; and that south of the river, known as 'fielden Warwickshire', relatively wealthy, well-cultivated and with little woodland. Moreover, certainly by the thirteenth century, no part of the county was subject to the forest law. It is not surprising, therefore, that it supported some 52 parks, of which all but three lay north of the Avon.

The close correlation between the existence of Domesday woodland and subsequent imparking is also apparent in the case of Leicestershire and Buckinghamshire.[51] In the former county, the woodland was disposed in two main areas, in the western half and in the south-east. That in the west roughly coincided with the Charnwood Forest, which though not a legal forest contained substantial amounts of waste and woodland, and that in the south-east with the Leicestershire portion of the Forest of Leicestershire and Rutland, was disafforested in 1235.[52] It is precisely in these two sections of the county that the great majority of its parks are located, while such areas of fertile clay, agricultural land as the Welland valley and the Vale of Belvoir, which had been cleared of woodland before the eleventh century, were virtually devoid of parks. In Buckinghamshire, a similar pattern emerges and the central part of the county, the pastoral and arable Vale of Aylesbury, which had lost much of its woodland cover by 1086, contained very few parks, while the wooded areas to the north and the south of the county were thickly populated with parks. It would seem very probable, therefore, that, as in the case of moated homesteads, the distribution of parks was to a considerable extent a response to economic and geographical circumstances whereby large landowners with numerous manors would impark only on the convenient ones where imparkment would not impinge upon cultivation, while small landowners, with perhaps only a single manor, rarely imparked in intensively cultivated districts.

Legal restraints, such as the forest law, also inhibited imparkment and royal forests generally contained few parks. As we have seen, they were notably absent in the Staffordshire forests of Cannock and Kinver and the Dorset forests of Blackmore and Purbeck, and the same was true in Gloucestershire, where the Forest of Dean contained very few parks, and in Wiltshire, where the forest law was enforced throughout the Middle Ages and where the density of imparkment remained relatively low. In some forests, however, the king did create parks and occasionally granted the same right to his subjects: in Windsor Forest, for example, there were 15 parks, all but two of which belonged to the king.

The impact of the medieval park on the landscape was clearly a considerable one, which is hardly surprising given that there were at least 1,900 parks in existence at various times in the Middle Ages for which definite evidence exists, and probably many others which have left no documentary or other trace. Indeed, their impact on the landscape is, in many places, still discernible today in the form of parts of the earth bank which constituted the basic boundary of the park, field-names and farm-names, and curving hedge lines which mark the line of former park boundaries. Regrettably, as with other intimate features of the English landscape, modern farming methods have resulted in the destruction of hedges and boundary banks in order to create larger fields.

Warrens

During the Middle Ages, the term *warren* was used in two quite different ways. The first, with which we are chiefly concerned because of its imprint on the landscape, described a rabbit warren, as we understand it today, namely a place where rabbits are preserved and encouraged to breed. The second was used to define the exclusive right to hunt in a specific place, that is the right of warren, or *free warren*. This was a privilege granted by the Crown to local landowners enabling them to hunt smaller game — the fox, rabbit, hare, wild cat, badger, marten, otter and squirrel and pheasant and partridges — over their estates. The right of free warren therefore meant both the right to hunt particular warrenable animals and also the place where the right was exercised.[53] According to some authorities, the hare was the principal beast of the warren giving the best sport when hunted fairly with hounds.[54] The rabbit, by contrast, was not much hunted for sport but was highly regarded for its meat and skin. Unlike the hare, it was not native to this country but was probably introduced into England in the twelfth century.[55] The other beasts of the warren such as the fox, wild cat, badger, marten, otter and squirrel were hunted mainly because they were regarded as harmful to deer, crops or domestic animals. Rights of free warren were granted from the Conquest onwards by William I and William Rufus, and both they and later kings frequently granted this privilege to monasteries situated within royal forests.[56] By the middle of the fourteenth century, grants of free warren had become so common that the majority of manorial lords seem to have enjoyed them; indeed, in some counties, like Rutland, these rights continued well into the modern period.[57]

The right of free warren, therefore, constituted a general permission to hunt specific animals over manorial lands and contributed little or nothing to the making of medieval landscape. The *warren*, by contrast, was a distinctive feature of the landscape and was an enclosed area of land varying in size from a relatively small field to a square mile or more, used for breeding rabbits. It belonged to the lord of the manor who, in the case of a larger warren, might employ a warrener to look after it: for example, in 1300 the Earl of Cornwall paid his warrener at his manor of Oakham 6s 5d a year.[58] As rabbits were also known as 'coneys' or 'conies', by extension warrens were also known as coney-garths or coneries, or, in the case of one within Leicester Forest which was rather more than two acres in extent, a 'coninger'.[59]

The first references to rabbits in England occur in 1176 when they were found in the Scilly Isles.[60] However, they were still probably very scarce at this time and the first certain mention of native mainland rabbits was in 1235 when Henry III made a gift of ten live rabbits from his park in Guildford; six years later, in 1241, the first reference occurs to a coneygarth on the English mainland, in the same park.[61] During the next decade or so, warrens multiplied and in 1268, for example, Richard, Earl of Cornwall complained that his coneygarth at Isleworth in Middlesex, was being plundered by poachers. The most common way of capturing them seems to have been with hawks, dogs and ferrets.[62]

The medieval warrens were quite numerous and in various places have left their mark on the contemporary landscape. In Ashdown Forest in Sussex, for example, where they are known as 'berrys', they were enclosed with perimeter banks and ditches and contained earth banks in the form of long narrow 'pillow mounds' for the rabbit burrows.[63] In addition to the earthworks thrown up to create them, rabbit warrens had another effect on the appearance of the countryside, in that the practice of rabbit warrening often prevented the regeneration of trees and encouraged the growth of bracken.[64] Doubtless, the warrens would from time to time get out of hand as in the case of Leicester Forest in 1605, where a jury of commoners complained that three warrens had 'overspread and fed over a hundred acres of ground and more, to the oppression of the commoners and the utter exile of his Majesty's game'.[65] Finally, the medieval rabbit warren has left its mark in the names of field, hill and farm as many a 'Warren Farm' or 'Conegar Hill' testify.

Notes

1. S.A. Moorhouse, 'Documentary Evidence for the Landscape of the Manor of Wakefield During the Middle Ages', *Landscape History*, vol. 1 (1979), p. 51.

2. H.C. Darby, *The Domesday Geography of Eastern England* (Cambridge University Press, Cambridge, 1952), pp. 56, 124, 126, 182, 234-5, 330, 335.

3. C.R. Young, *The Royal Forests of Medieval England* (Leicester University Press, Leicester, 1979), p. 2; and J.R. Birrell, 'The English Medieval Forest', *Journal of Forest History*, vol. 24, no. 2 (April 1980), p. 78.

4. O. Rackham, *Ancient Woodland* (Edward Arnold, London, 1980), pp. 175-9.

5. H.C. Darby, *A New Historical Geography of England before 1600*, (Cambridge University Press, Cambridge, 1976), p. 55.

6. H.C. Darby, *Domesday England* (Cambridge University Press, Cambridge, 1977), includes a map of royal forests in 1086 (p. 197) and an appendix listing references to forest (pp. 354–5).

7. Young, *The Royal Forests of Medieval England*, p. 6.

8. Ibid., p. 170.

9. R. Cunliffe Shaw, *The Royal Forest of Lancaster* (The Guardian Press, Preston, 1956), p. 90.

10. G.H. Cook, *English Monasteries in the Middle Ages* (Phoenix House, London, 1961), p. 28.

11. For the Forest of Dean, see M. Bazeley, 'The Forest of Dean in its Relations with the Crown during the Twelfth and Thirteenth Centuries', *Trans. Bristol and Glos. Arch. Soc.*, vol. 33 (1910), pp. 191–202; and for Savernake Forest, see the Earl of Cardigan, *The Wardens of Savernake Forest* (Routledge and Kegan Paul, London, 1949); and John Rodgers, *The English Woodland* (Batsford, London, 1942), pp. 63–4.

12. Young, *The Royal Forests of Medieval England*, p. 11.

13. C. Chenevix Trench, *The Poacher and the Squire* (Longman, Harlow, 1967), p. 24.

14. H.L. Edlin, *Trees, Woods and Man* (New Naturalist Series), 3rd edn (Collins, Glasgow, 1970), pp. 111, 177–80; and O. Rackham, *Trees and Woodland in the British Landscape* (Dent, London, 1976), p. 74. An important new book which describes in detail the medieval forests, including the trees and animals, the administration of the forest laws, and descriptions of individual forests and chases is N.D. James, *A History of English Forestry* (Basil Blackwell, Oxford, 1981).

15. Rackham, *Trees and Woodland in the British Landscape*, p. 71.

16. Ibid., pp. 72–4.

17. For detailed examinations of the economic administration of specific forests, see Bazeley, 'The Forest of Dean in its Relations with the Crown during the Twelfth and Thirteenth Centuries'; J.R. Birrell, 'The Forest Economy of the Honour of Tutbury in the Fourteenth and Fifteenth Centuries', *Univ. Birmingham, Hist. J.*, vol. VIII (1962), pp. 114–34; and the Earl of Cardigan, *The Wardens of Savernake Forest*.

18. Young, *The Royal Forests of Medieval England*, p. 131.

19. Birrell, 'The English Medieval Forest, pp. 81–3.

20. Young, *The Royal Forests of Medieval England*, pp. 130–4.

21. B.E. Harris (ed.), *Victoria County History, Gloucestershire*, vol. 2 (OUP, Oxford, 1979), pp. 167–87.

22. L.M. Cantor, 'The Medieval Forests and Chases of Staffordshire', *N. Staffs. Journal of Field Studies*, vol. 8 (1968), p. 46.

23. Rackham, *Trees and Woodland in the British Landscape*, p. 156.

24. W.J. Liddell, 'The Private Forests of S.W. Cumberland', *Trans. Cumb. and West. Antiq. and Arch. Soc.*, vol. LXVI, new ser. (1966), p. 106.

25. Chenevix Trench, *The Poacher and the Squire*, p. 37.

26. L. Fox and P. Russell, *Leicester Forest* (Edgar Backus, Leciester, 1948), p. 48.

27. Cunliffe Shaw, *The Royal Forest of Lancaster*, p. 6; and Cantor 'The Medieval Forests and Chases of Staffordshire', p. 49.

28. Fox and Russell, *Leicester Forest*, pp. 31–43.

29. Cantor, 'The Medieval Forests and Chases of Staffordshire', pp. 49–50.

30. Ibid., p. 50.

31. C.R. Elrington and N.M. Herbert (eds.), *Victoria County History, Gloucestershire*, vol. 10 (OUP, Oxford, 1972), p. 51.

32. J. Gould, *Men of Aldridge* (G.J. Clark, Walsall, 1957), p. 34.
33. Cantor, 'The Medieval Forests and Chases of Staffordshire', p. 48.
34. D. Hawkins, *Cranborne Chase* (Victor Gollancz, London, 1980), pp. 31–5.
35. This section draws heavily on L.M. Cantor and J. Hatherly, 'The Medieval Parks of England', *Geography*, vol. 64 (1979), pp. 71–85, which contains a full bibliography on the subject.
36. For a detailed description of Moulton Park, see J. Best, 'A perambulation of Moulton Park', *Confluence* (Nene College, Northampton, September 1979), pp. 17–40.
37. J. Manwood, *A Treatise of the Laws of the Forest* (1615), p. 18.
38. G.J. Turner (ed.), *Select Pleas of the Forest* (Selden Soc., XIII, London, 1899), pp. cxiv ff.
39. L.M. Cantor and J. Hatherly, 'The Medieval Parks of Buckinghamshire', *Records of Bucks*, vol. XX (1977), pp. 442–3.
40. Darby, *Domesday England*, p. 201.
41. Ibid.; see especially the map and list of Domesday parks, pp. 202–3.
42. For a description and detailed perambulation of Gillingham Park, see L.M. Cantor and J.D. Wilson, 'The Medieval Parks of Dorset, V', *Proc. Dorset Nat. Hist. and Arch. Soc.*, vol. 87 (1966), pp. 1–5.
43. Cantor and Hatherly, 'The Medieval Parks of England', p. 78.
44. C. Platt, *Medieval England* (Routledge and Kegan Paul, London, 1978), p. 47.
45. L.M. Cantor and J.S. Moore, 'The Medieval Parks of the Earls of Stafford at Madeley', *N. Staffs. Journal of Field Studies*, vol. 3 (1963), pp. 37–58.
46. Rackham, *Trees and Woodland in the British Landscape*, p. 145.
47. M. Aston and T. Rowley, *Landscape Archaeology* (David and Charles, Newton Abbot, 1974), pp. 154–5.
48. Royal Commission on Historical Monuments (England), *An Inventory of the Historical Monuments in the County of Northampton, Vol. II, Archaeological Sites in Central Northamptonshire* (HMSO, London, 1979), pp. lvii–lix.
49. B.K. Roberts, 'Medieval Fishponds', *Amateur Historian*, vol. 7 (1966), pp. 119–25.
50. L.M. Cantor and J.D. Wilson, 'The Medieval Deer-Parks of Dorset, I', *Proc. Dorset Nat. Hist. and Arch. Soc.*, vol. 83 (1961), pp. 109–11.
51. See L.M. Cantor, 'The Medieval Parks of Leicestershire', *Trans. Leics. Arch. and Hist. Soc.*, vol. 66 (1970–1), pp. 9–24; Cantor and Hatherly, 'The Medieval Parks of Buckinghamshire'.
52. L.M. Cantor, 'The Medieval Hunting Grounds of Rutland', *Rutland Record*, (Journal of Rutland Record Society), (1980), pp. 13–18; H.C. Darby and I.B. Terrett, *The Domesday Geography of Midland England*, 2nd edn (Cambridge University Press, Cambridge, 1971), p. 344.
53. Hawkins, *Cranborne Chase*, p. 53.
54. Chenevix Trench, *The Poacher and the Squire*, p. 38.
55. C. Lever, *The Naturalised Animals of the British Isles* (Paladin, St Albans, 1977), p. 64.
56. Young, *The Royal Forests of Medieval England*, pp. 10, 44.
57. Cantor, 'The Medieval Hunting Grounds of Rutland', p. 14.
58. W. Page (ed.), *Victoria County History, Rutland*, vol. 1 (Constable, London, 1906), p. 13.
59. Fox and Russell, *Leicester Forest*, p. 84.
60. Lever, *The Naturalised Animals of the British Isles*, p. 65.
61. Ibid., p. 66.
62. Ibid., p. 69.
63. P. Brandon, *The Sussex Landscape* (Hodder and Stoughton, London, 1974), pp. 110–11.
64. Ibid., p. 156.
65. Fox and Russell, *Leicester Forest*, p. 96.

4 MARSHLAND AND WASTE

Michael Williams

By the time of Domesday, much of the landscape of lowland England had been settled and colonised; in some ways it was already an 'old' landscape. Most of the villages that we recognise today were created by 1086. Around them the large, arable open fields stretched away for perhaps half a mile or a little more, but rarely reached the edge of the village territory — often the ecclesiastical parish boundary — because that was still uncolonised marsh, and scrub and woodland. These were usually called 'waste', but in reality they were a valuable resource; the first provided common grazing and occasional hay, sedges and rushes for thatching and flooring; and the second was useful for pannage, for constructional timber, and was absolutely indispensable for fuel. On the basis of the evidence available, it seems probable that perhaps one-half of the village territory at this time still remained to be reclaimed. In the north and the west of the country, where there were extensive areas of rocky ground, thin soils, steep slopes and woodland, the compact, communally cultivated village settlements gave way to smaller, clustered settlements and individual houses, surrounded by small irregular fields that had been won from the waste bit by bit.

After Domesday, although the expansion of settlement produced few new villages it did produce a large number of new farms and fields. The new expansion of population (by perhaps between twofold and threefold) was accompanied by an extension of farmland, so that the eleventh to early fourteenth centuries were a period of agricultural colonisation.[1] The large fields of the existing villages reached out towards their boundaries as improved techniques, additional ploughs and more people meant that the existing land was not only exploited more intensively, but additional land was taken in from the peripheral 'waste', villagers nibbling away piecemeal at small parcels of wood, heath, moor and marsh. The village lands were added to furlong by furlong, each strip being a physical manifestation of the labour of a limited group of men over a limited time, intent on getting land into cultivation quickly. In other places, the colonisation of the waste was more spectacular because large areas of hitherto unused or under-used moorland, marshland, heathland and woodland were being reclaimed and brought into use. There was still room to expand (Figure 4.1):

marshland covered most of the Fens and the Somerset Levels; woodland covered the Arden and Feckenham areas of the north-west Midlands, and much of Wealden Kent and Sussex, as well as the interfluvial areas everywhere else; while large areas in Devon, Hampshire, Berkshire, Essex and Northamptonshire, to mention but a few counties, were under royal forest law. On the edge of the upland in the west and the north, there was land for those with the will and energy to work, and in the 'wasted' areas of the north, derelict land and abandoned land were available for reclaiming.

Beresford and St Joseph in their admirable aerial photographic and literary survey of *Medieval England* have suggested that for much of the population life consisted of 'the journey to the margin'.[2] But it is important to recognise that the margin was of a twofold nature; not only was it the land on the edge of the existing settlement, as is commonly supposed, but it was also the least rewarding and the most 'difficult-to-reclaim' land wherever it might have been located. The return from it was sometimes barely worth the labour expended, so that it was economically marginal land even in the midst of well-settled areas.

Often the work of reclamation was that of large landholders, usually ecclesiastical estate holders, but freeholders were also active, far more commonly than is realised. Where a lord had areas of waste on his domain he was disposed to offer concessions of freedom from feudal labour services to attract colonists who would do the reclamation work, thereby increasing his rent roll. It would be wrong to think of large landholders and small freeholders as mutually exclusive categories, however, as in many places they are difficult to separate. For example, the massive reclamation work in the silt lands of the Fens was primarily the work of numerous freeholders working co-operatively, and only later did the great ecclesiastical houses organise reclamation. This mix of individuals and institutions pushing out the frontiers of cultivation was present in most localities in medieval England, so that the quest for profits by the large landowners and the quest for survival by the peasants and freeholders went hand in hand.

The physical expansion of settlement and cultivation between the Conquest and the Black Death was, perhaps, the notable feature of the age, and a central theme in the economic development of the country. Crucial to that expansion was the growth of the population of England, whch may have been between 1.25 and 2.25 million (probably well into the upper part of that range) in 1086, and could have risen to as great as 4.0 or 4.5 million in 1377. But if we consider the devastation of the

Figure 4.1: The major regions and places of wasteland reclamation mentioned in the text

plagues of 1348-9, 1360-2 and 1375, when probably between 30 and 50 per cent of the population was wiped out, then it is possible that the population just prior to the Black Death of 1348 could have been as much as 5 or 6 million. The accuracy ascribed to these absolute figures is open to lively debate, particularly the base for Domesday against which all increase is measured, but there is little dispute among historians that the population must have at least doubled or even tripled between the late eleventh and early fourteenth centuries, leading to a near Malthusian crisis.[3]

Given the expansion of the numbers of people and the expansion of the area of cultivation during the early Middle Ages it would be reasonable to ask which was cause and which effect. Did the improved food supply result in a healthier population which lived longer and had less infant mortality (aided undoubtedly by more peaceful conditions which obtained for much of the period, and the decline of epidemic diseases), or did the awareness of the surplus land and the possibility of upgrading it from waste to some higher form of productivity both encourage and allow people at subsistence level to have larger families? A third possibility is that suggested by Duby, that technical progress and expansion 'never succeeded in meeting the needs of a teeming population which lay at the mercy of a shortage of food as cruel, perhaps, as it had ever been in Carolingian times'.[4] It is a problem of cause and effect that baffles economists in contemporary situations, so the chance of answering the question satisfactorily for the medieval period is remote. One can say, however, that there is much evidence of over-population, ranging from declining yields to the abandonment of land as evidenced in the *Nonarum Inquisitiones* of 1341.[5] In addition, it seems clear that the greatest known rates of population growth took place in those areas which had the greatest potential for development, that is with the greatest expanse of 'waste'. For example, in the Lincolnshire Fenland the number of recorded households increased by more than sixfold at Spalding, and as much as elevenfold at Pinchbeck between 1086 and 1287, and this trend continued in successive decades. Some completely new settlements such as Fleet recorded increases of over sixtyfold, and the whole area around the Wash, particularly in Holland and the Norfolk marshland, was the richest area of England in 1334.[6] In the wooded areas of Arden in Warwickshire, increases of population ranging between four and eightfold were recorded between 1086 and 1279.[7] Both in the Wash and Arden adjacent areas with little potential for development hardly grew at all. Again, the estates in the border counties with Wales and Scotland, and particularly in Yorkshire, devastated by the progress of the Norman armies, and Welsh and Scottish raiders at various times from the mid-twelfth century onwards, subsequently recorded major upswings of population as the 'wasted' land was brought back into production. In contrast, the block of counties in the southern half of the country appeared to be growing only very slowly, although, of course, from higher base populations.[8]

The implication is, then, that the expansion of population was a response to the ability to expand the land under cultivation, and not vice versa. However, one cannot be too dogmatic about this, as in so many matters related to medieval England, since it was probably a two-way process, the presence of more land capable of development, meaning more food, allowing more people to live, and the pressures of more people, as generations multiplied, creating a demand for more intensive and extensive use of hitherto idle land.[9] To our knowledge, food supplies at least did not act as a brake on population increase.

Tracing Colonisation and Reclamation

Because the colonisation and reclamation took place between 850 and 650 years ago in a largely pre-literate age, where few records were kept and from which even fewer records have survived, there is much we do not know. Consequently, the identification of the areas affected depends upon two different scales and sorts of analysis. First, the reduction in the area of waste may be adduced by comparing the distribution patterns derived from national records that contain numerical data of different types of wealth at widely spaced intervals, for example Domesday values and the Lay Subsidies. Another wide-ranging approach has been to classify the counties according to predominant population, agricultural and reclamation trends. Second, the reduction in the area of the waste may be adduced from a painstaking reconstruction of local events, almost field by field, say, within a parish, from the evidence of custumals, rent rolls, charters, manorial surveys, field-names, and the like.

At one extreme we run the risk of gross generalisation, inaccuracy and the formulation of many assumptions; at the other extreme we run the risk of losing ourselves in a morass of detail which may be purely local in its occurrence and limited in its applicability. The middle ground seems achieved only occasionally, where, for example, local detail is coupled with wider regional information, as in those areas that were included in ecclesiastical estates for which copious records exist, or where the process of colonisation has left definite and distinctive physical vestiges in the landscape, such as baulks, ditches, walls, buildings and roads, together with place-names, that can be seen, dated and interpreted. This is particularly true of marshland areas where new land was created and made. Of these and similar areas we can agree with W.G. Hoskins, that

> There are certain sheets of the one-inch Ordnance Survey maps which one can sit down and read like a book for an hour on end, with growing pleasure and imaginative excitement . . . One dissects such a map mentally, piece by piece, and in doing so learns a good deal of local history, whether or not one knows the country itself.[10]

Consequently, one must take cognisance of both approaches in an attempt to arrive at valid conclusions as to the location, timing and appearance of the landscape of medieval colonisation and reclamation. However, the limitations of the methods must be recognised and appreciated if we are to treat the data in a scholarly way.

The first scale of analysis is based on the reasonable premiss that any reduction in the amount of waste land would lead to an increase in the material wealth of the area affected as more people settled there, production increased, and the land values and taxable base rose. Both Buckatzsch and Schofield have shown conclusively that wealth rose in medival England,[11] but both their studies were based on county units the size of which tends to obscure the finer variations that we have hinted at already. Since then, H.C. Darby *et al.*,[12] have supplied an overview of changes, not in terms of *absolute* wealth but in terms of *relative* prosperity, one area compared with another, between the Domesday Inquest of 1086 and the Lay Subsidies of 1334 and 1524/5 plotted for 610 units. The tax of 1334 occurred at approximately the end of the years of expansion 'and before the full impact of the recession of the later Middle Ages was felt'; the 1525 subsidy came 'after the recovery that took place towards the close of the fifteenth century and in the early years of the sixteenth'.[13]

There are problems in utilising and comparing these sets of regional indicators of wealth, one important deficiency being that the 1334 subsidy excluded church property, as well as the wealth of the Cinque ports and the Stannary men (tin miners) of Devon and Cornwall. In 1525, there was significant under-enumeration of the estates in the

Figure 4.2: The distribution of wealth in medieval England: the highest quintiles, 1086–1524/5

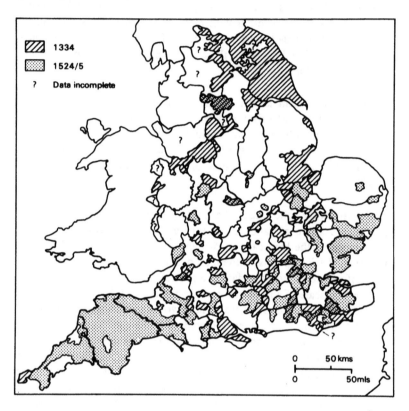

Note: 1334 refers to the 1334 assessed wealth per Domesday man (£3.25 and over). 1524/5 refers to the tax of 1524/5 as a percentage of 1334 assessed wealth (16.65 per cent and over).
Source: Based on H.C. Darby *et al.*, 'The Changing Geographical Distribution of Wealth in England, 1086 – 1334 – 1535', *Journal of Historical Geography*, vol. 5 (1979).

north.[14] However, despite these limitations the comparison between areas of greatest change (the highest quintiles of the distributions mapped) suggest radical changes in parts of the country between 1086 and 1334 which were very different from the distribution of changes between 1334 and 1525 (Figure 4.2).

Hallam has suggested a more intuitive classification 'for assarting from the waste between the fifth and fifteenth century', based on an examination of place-name evidence occurring either side of Domesday

Figure 4.3: Assarting from the waste between the fifth and fifteenth centuries

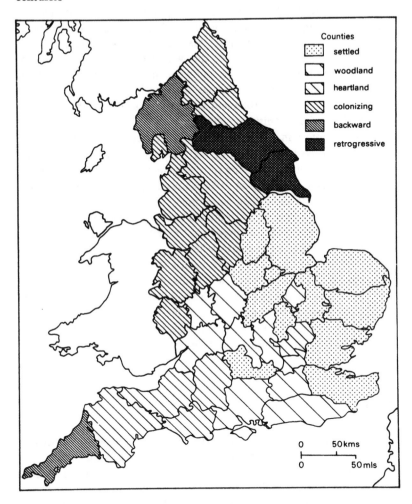

Note: See text for explanation of categories.
Source: Based on H.E. Hallam, 'The Postan Thesis', *Historical Studies*, vol. 15 (1972).

and a detailed consideration of the known population and agricultural trends in different parts of the country.[15] He suggests that the counties of England can be divided into six categories, which he has labelled settled, woodland, heartland, colonising, backward and retrogressive (Figure 4.3). The settled counties 'developed very early very far and went little further after 1086'; the woodland counties 'developed early, were well settled in 1086 and went much further in settlement after 1086'; the heartland counties were 'well developed before the English Conquest and in 1086 and showed few signs of further progress at any time'; the colonising counties 'were poorly developed in 1086, started to develop in mid-Saxon times or later and developed fast and far later'; the backward counties 'were poorly developed at all times and did little about it'; and the retrogressive counties 'developed early but were poorly developed in 1086, showed signs of regression before 1086 and did little about their condition'.

By looking at these broad frameworks it is possible to gain a generalised view of the changes taking place. Nevertheless, it is inevitable that one continually looks for confirmation of these changes by examining parts of the country at the other, local and detailed scale of analysis. Consequently, the marshland, the woodland and the uplands, as well as the special case of the granges which straddled all types of reclamation, will be looked at in turn. In many cases the detailed study backs up the generalised conclusion, but in other cases it does not. Our knowledge of the transformation of the waste in medieval times is still far from perfect.

Marshland Colonisation

Of all the landscapes created by the reclamation of the waste during the medieval period that of the marshlands was the most spectacular. New lands appeared where none existed before, swamps became dry land, and old drainage channels were regraded and reoriented, and new ones dug. Along the coast, walls and groynes kept out the tidal surges and hastened the deposition of silt.

In addition to these physical changes, new societies and new economies were established as human endeavour and ingenuity combined with the co-operative enterprise of individuals to create new geographies in the coastal areas of England (Figure 4.1). The Fens in eastern England and the Levels of central Somerset in the west of England, were two of the largest areas affected, each on opposite sides of the country, each displaying significant differences but also many similarities. But there were other areas. To the south were the Norfolk Broads, the Essex Marshes, Romney Marsh in Kent and Pevensey Levels in Sussex; to the north there was the Isle of Axeholme, the Hull Valley, the Vale of York, the Humberhead Levels, and the Lancashire Mosses, to name the most important. Almost all of these lowland regions had a distinctive individuality which is reflected still in their descriptive regional names like Fens, Levels, Carrs, Mosses and Broads.

Because the evidence of Domesday shows that the settlements of the time carefully avoided the lowlands and clung to the flood-free uplands and islands, there is a tendency to think of these lowlands as extensive swampy morasses, bereft of settlement and devoid of use. But nothing could be further from the truth. There were large areas of swampy ground, sometimes under water for years on end, but more often than not for a limited period only during the autumn and winter, their deepest parts remaining as standing water in pools and meres. But for most of the time the lowlands were not unproductive, desolate and dangerous, but extremely useful ground, technically part of the manorial waste, to be sure, but valuable ground zealously guarded by the surrounding settlements on the uplands. Evidence abounds from the time of Domesday to the fourteenth century of the traditional occupations and products of the marshlands and can be summarised by suggesting that there was almost a hierarchy of usefulness in the low grounds that depended on the state of the drainage. Nearly every watercourse and pool had its fishing weirs (*gurgites*) and fisheries; eel renders reached thousands and even tens of thousands per annum in scores of places in the Levels and the Fens, and fish and eels fulfilled many of the uses of currency as debts, payments, and rents were often paid in them. Just as fish were a part of a varied economy and, perhaps even more important, a varied diet, so were fowl, but unfortunately this was more a haphazard, chancy exploitation dependent on the vagaries of bird migration habits, and although there are many hints about their value, there is no great body of evidence to show how important they were.[16] Salt pans, and rush and sedge collection were important locally, and turbary is mentioned increasingly from about 1200, when large areas of peat fen were reserved for digging for fuel, the supply of wood being very limited in the marshlands. The record is sketchy in places, but in the Norfolk Broads it is now very clear. Archaeology, palaeobotany, and the evidence of documentary and sedimentary research come together to show quite conclusively that beneath the open water of the Broads as we know them today are drowned medieval peat diggings worked mainly by St Benet's Priory. If the volume of the present lakes of the Broads is calculated then at least nine billion cubic feet of material must have been removed before the water/land levels changed relative to one another in the early fourteenth century and drowned the diggings. This amount of extraction is not beyond the bounds of possibility, given the substantial amounts known to have been taken out each year and the number of peat-digging communities who were active over a period of three hundred years.[17]

But it was as common grazing that the marshlands were most used and most valuable. Pasture rights soon became an integral part of the manorial economy as groups of villages agreed to demarcate parts of the lowlands for their exclusive use. The custom varied: many of the villagers were allowed to pasture as many animals as they liked — *common sans nombre*, as many of the documents called it; some villagers limited the number of beasts; and outsiders usually paid a fee. Whatever the custom, in many places the rights tended to pass to the lord, and in time the villagers paid for the right to common. The policing of these customs was made possible by having droves from time to time to round up the cattle, the offenders getting their beasts out of the pound only by payment of a fine. Some groups of villages and individual landholders began to subdivide the lowlands further in an attempt to protect their rights, and almost inevitably as grazing became more regulated disputes between individuals and institutions became more frequent. These were often very violent and took the form of battles of arms, the destruction of ditches and marker stones, the killing and maiming of animals, the pulling down of buildings and, in Somerset, the firing of the combustible peat moors. The disputes between Crowland Abbey and the surrounding village communities have been documented fully, and the tripartite duel between the Bishop of Bath and Wells, the Dean and Chapter of Wells, and the abbots of Glastonbury for the mastery of the Brue Valley pastures make equally fascinating reading. These disputes are full of details about the usefulness and value of the moors, and it is not hard to believe that there were many other disputes which have passed out of memory but which may have been equally revealing about the medieval marshlands.[18] Nevertheless, the fact remains that the delimitation of intercommoning rights and the division of the marshlands occurred during the thirteenth and early fourteenth centuries, and the bitterness of the disagreements underlines the vital position occupied by pasture rights in the economy of these marshlands. The division had an even greater significance than that, however, for taken one step further it led inevitably to the enclosure and reclamation of the marshland and the transformation of the landscape.

Taken as a whole, the evidence suggests that the practice of inter-commoning on the open moors and fens was unsatisfactory; the un-limited rights and stocking arrangements were open to abuse, and both the length of the grazing season and the value of the natural pasture were uncertain because of flooding. Therefore, in order to ensure the pasture against human and natural depredations reclamations were made in the commons, a trend which was encouraged by the general division of the waste at this time and by the growing concern to obtain specific rights in more limited areas. In the creation of new meadows and in the extension of the grazing season lay the best means of enlarg-ing the economy, because winter fodder was crucial to medieval live-stock farming. An acre of pasture land converted into meadow could equal or more usually exceed in worth a comparable acre of arable land. This reclamation to make meadows was particularly appropriate in the environment of the Somerset Levels where annual landward flooding was an assured event, and arable farming had to be restricted to the upland locations. While flooding also occurred in the Fens the superior soils of the silt land, the lesser susceptibility to flooding, and the generally drier climate meant that reclamation was aimed at creating new arable farming land in the erstwhile waste. The building of walls and embankments, both outwards towards the sea and inwards towards the lower fens and moors meant that each successive intake widened the area of permanent settlement and reduced the area of the waste.

The Fenland

Perhaps the most striking example of reclamation in the marsh was in the belt of silt and adjacent peat fen that surrounds part of the Wash (Figure 4.4). In 1086 the silt belt was very much poorer than the surrounding uplands; by 1334 it was many times richer.[19] The progress of reclamation and the changes in the status of the marsh have been painstakingly reconstructed by Hallam from thousands of documents for the Lincolnshire portion of the Fenland,[20] and although no one has yet done the same for the Norfolk and Cambridgeshire portions on the southern side of the Wash, it is reasonable to assume that the same held true there.[21]

The silt lay several feet above the level of the inland peat fen and the seaward marsh, and the settlement of this largely flood-free belt of land occurred from the seventh century onwards with notable Danish additions. The extent of that settlement is easily and reasonably

Figure 4.4: Reclamation in the medieval Fenland

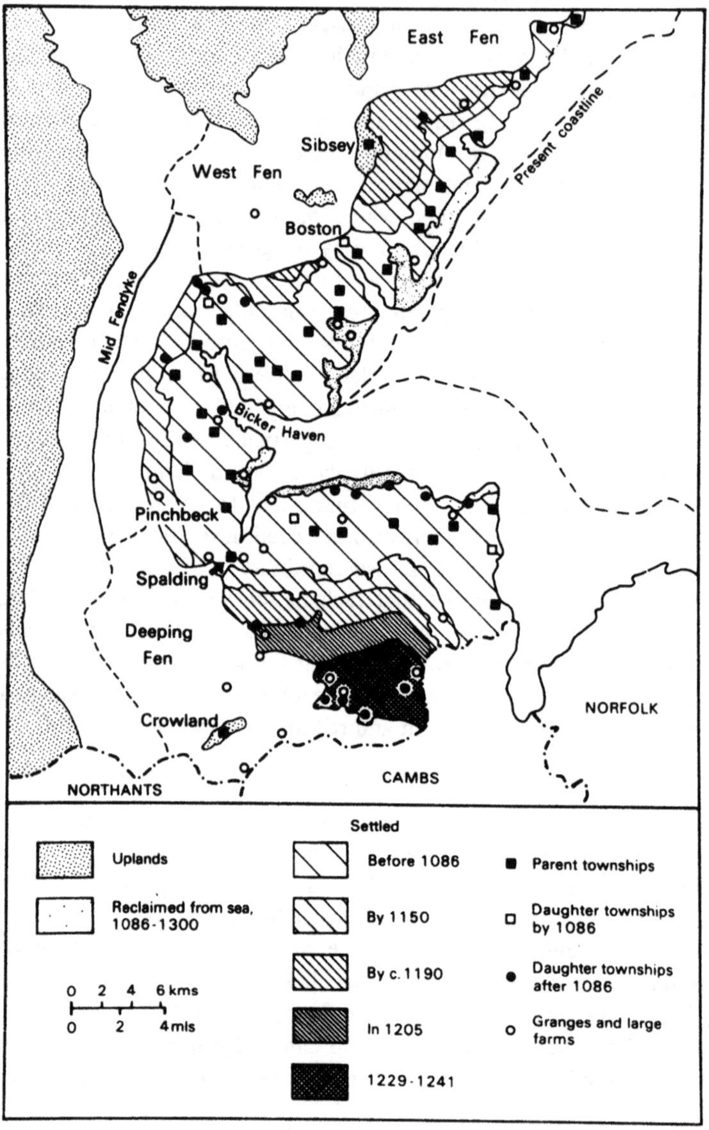

Source: Based on H.E. Hallam, *Settlement and Society: A Study of the Early Agrarian History of South Lincolnshire* (Cambridge University Press, Cambridge, 1965), by permission of the Cambridge University Press.

accurately determined by a perusal of the relevant portion of the
Ordnance Survey map of any part of the belt. The creation of this
'map' derives in large part from the fact that between about 1150 and
1300 the ordinary men of the villages around the Wash engaged in a
massive feat of reclamation that won some 16 square miles from the
marshland bordering the Wash and some 90 square miles from the fenland. The reclaimed land was used mainly for arable farming to feed
what seemed to be a rapidly growing population. The evidence of their
bustling youthfulness and energy in colonisation of the waste is etched
indelibly on the landscape by a series of walls or dykes, and also by
many additions, each dyke marking a step in the reclamation process.

The chronology of colonisation is most clearly portrayed in the
wapentake of Elloe, east of Spalding.[22] On the seaward side of the silt
belt reclamations were being made in estuarine land, each 'inning'
being smaller than those pieces of reclaimed inland, as we shall see,
but significant both because they were generally earlier and also because they represented an addition to the sea defences, deposition
behind the sea walls often resulting in ground being at a higher level
than the land in the silt belt. Nature taught the Fenmen what to do;
the evidence before their eyes of siltation of the estuaries and the
gradual rise of the salt marshes showed them that groynes, walls and
indeed any impediment to the natural flow of the silt-laden water was
an obvious way of literally making the new ground which they needed.
The pieces of reclaimed land almost invariably had 'new land' incorporated in their name, for that is exactly what they were. Although all
drainage works needed upkeep and maintenance, the most important
were the sea banks around the new lands. It was imperative to keep the
sea walls in good repair against the onslaught of high tides and sudden
storms, and consequently maintenance was rigidly observed, so many
feet of wall being the responsibility of the owner of so many acres of
land. Maintenance demanded an efficient dyke-reeve organisation with
a means to punish or coerce the slack landowner whose negligence
potentially endangered the property and lives of all his neighbours.
Also, the construction of sluices in the walls, to let out the trapped
water at low tide, showed an appreciation of the nature of the basic
drainage techniques. That all these were present was a measure of the
social and technical organisation of the Fenmen.

But it is on the Fenland side of the silt that the evidence of reclamation is so dramatic. Eastwards from Spalding lay the 'considerable
villages' of Weston, Moulton, Whaplode, Holbeach, Gedney, Fleet,
Sutton, Lutton, and Tydd St Mary. To the south of them is a series

of east–west trending fen banks, each marked today by a major east–west road. The first and earliest dyke is known variously as Austen Dyke, Hurdletree or Fen Dyke. It is probably of pre-Conquest origin and it is fourteen miles long. About one mile to the south is the line of Old Fen Dyke, Saturday Dyke or Raven's Dyke, constructed between 1160 and 1170; beyond that again is Hassock Dyke or Old Fen Dyke constructed between 1190 and 1198; beyond that again, Asgar, Lord's or New Fen Dyke constructed between 1203 and 1208; and out on the marsh edge there is Common Dyke probably built between 1206 and 1241. Thus in less than eighty years, some 50 square miles of new arable and pasture lands were added to these townships, the last great reclamation at the beginning of the thirteenth century adding at least 18 square miles of land.

From the disposition and alignment of the dykes it looks as though the villages co-operated in their construction, roughly in groups of three (hence the varying names for the dykes), although during the last two reclamations all the villages agreed first to divide the common fen equally between them, and then to co-operate in the building and maintenance of the new dykes. It was an impressive exercise in co-operative action and mutual aid, yet there was a paradox in this strong communal action, for the land was held in severalty and each owner could use it as he thought fit, which was the epitome of individualism. For these men of Elloe, or indeed for any of the silt lands of the other wapentakes of Lincolnshire, there appears to be no common field husbandry of the conventional Midland pattern.

The great new dykes were not the only distinctive elements in the landscape of the lowlands; it was crossed by north–south droves that allowed the villagers access to the common pastures and fields, and also by parallel ditches which helped drain the land and divide off the parish shares. These features remain today, the ditches often being parish boundaries, some parishes stretching for seventeen miles into the fen. Inevitably, too, the settlement pattern was modified as the immense distances between the parent villages and their new land necessitated some dispersal of residences. In Elloe, the Fen hamlets of Cowbit and Moulton Chapel grew up in the fourth enclosure of 1203–8, but it was not until the fifth and final bank was finished in about 1241 that distances necessitated the creation of the large number of hamlets of Whaplode Drove, Holbeach Drove, Gedney Hill and Sutton St Edmund. Along the coast, five daughter settlements were established in the new lands. The ecclesiastical houses, which had been partners in reclamation, adopted a similar practice. Crowland Abbey, for example, was surrounded by a ring of granges in its new enclosures, such as Ashwick Grange founded *c.* 1241, and Holland Fenhall, founded perhaps a little later between 1254 and 1271. Spalding Priory had Garrock Grange founded in 1285 and Goll Grange founded between 1298 and 1318.[23]

The process of reclamation along the sea coast and inland along the fen edge, described in detail for Elloe, was repeated in Kirton and Skirbeck wapentakes to the north, although our knowledge of the chronology of reclamation here is less certain because fewer documents are extant. Along the coast of Kirton, reclamation was less extensive than in Elloe, but the task was no less painstaking and important for the villages concerned. Inland, two parallel Fen walls, called the Old Fendyke and New Fendyke were constructed. They were situated at about one to one-and-a-half miles apart, trended northwards for about thirteen miles from Pinchbeck (actually in the westernmost corner of Elloe) to the head of Bicker Haven, the New Fendyke then turning eastwards for about six or seven miles to and near Boston on Boston Haven. The Old Fendyke was of unknown and remote origin, the New Fendyke was probably constructed between 1139 and 1180, although parts were much older, and between them lay the new enclosed lands of the villages of Pinchbeck, Surfleet, Gosberton, Church End, Donington and Bicker, while east of Bicker Haven and up to Boston Haven there were the new lands of Drayton, Wigtoft, Sutterton, Algakirk, Kirton, Frampton and Wyberton. In all, ten granges and large farms were established in the new lands, some by 'outside' religious houses, such as Sempringham Priory, but most by the silt-land settlements. There was little unused land, even outside the New Fendyke. The commons as far west as Middle Fendyke (a water course) were cut up after 1200 into numerous small and sporadic impermanent enclosures, both for crops and pasturing cattle.

North of Boston Haven lay Skirbeck wapentake. Again reclamation from the sea occurred, but it was more extensive than in Kirton, and amounted to perhaps ten square miles of new land, most coming into being between about 1140 and 1200. On the silt the villages of Skirbeck, Butterwick, Benington, Leverton, Old Leake, Sibsey, Wrangle, Friskney, Wainfleet St Mary and Wainfleet All Saints rapidly enclosed and reclaimed. The intakes between the Old Fendyke, or Ings Bank, and the New Fendyke were particularly rapid, occurring in less than a hundred years between the Conquest and 1150, which suggests a major pressure to expand. The fact that some enclosures were made by about 1180 beyond the New Fendyke towards the 'island' settlement of

Sibsey, and up to the very edge of the deep water-covered East Fen, suggests the same.

Finally, scores of old established villages along the northern and eastern uplands bordering the Fens were slowly incorporating into their parishes portions of the adjacent lowlands as far east as the Middle Fendyke. Sometimes the more favourable spots on the gravels were converted into arable cultivation, sometimes they were made into improved pasture for the taking of an occasional hay crop. If the peat was particularly well developed it provided valuable turbary beds, depending largely on the local state of the drainage. Sometimes these portions were known as Doles or Dales, being the shared ground of the settlement, most of which was apportioned during the mid-thirteenth century.

In addition to the sheer immensity of the task and the result of reclamation, there are perhaps other noteworthy aspects of the medieval Fenland reclamations. First, the environment of the Fens, with its products from the fields, fens, fresh waters and the seas provided a favourable means of sustaining a population which had already grown by 1086 to be many times larger than the average for other parts of Lincolnshire, and the country in general, and one must assume that the move to reclaim the 'waste' after about 1150 was a direct response to these large and growing numbers. Second, the reclamation was undertaken by whole communities of free peasants and smallholders, and the ecclesiastical houses did not get involved in reclamation until much later. For example, Goll Grange was created by Spalding nearly half a century after the enclosure of the fen in which it was situated. Therefore, the creation of granges and large farms in extensive activities in arable and pastoral farming was 'in fact a process not of colonization but of monopolization',[24] the monasteries engrossing and cashing in on the wealth created by the free peasantry. Third, the reclamation was intended primarily for arable farming, up to 70 per cent of the land in the Fenland being under crops in the late thirteenth century, and only 30 per cent in pastures and meadow.[25]

The Somerset Levels

The reclamation of the second largest area of lowland waste in the country, in the Levels of central Somerset, provides some interesting comparisons with the Fens. The higher rainfall, the greater tidal range in the Bristol Channel, and the susceptibility of the region to freak, heavy downpours that trigger off almost instant floods meant that farming, then as now, was much more likely to be pastoral than arable.

Second, by the early twelfth century much of the Levels was already under the control of the Abbeys of Muchelney, Athelney and Glaston-bury, and the Cathedral Church of Wells and they all took an active part in promoting reclamation on their estates, leaving little scope for reclamation by free men (Figure 4.5).[26]

On the other hand, the coastal clay belt of the Levels was like the silt belt of the Fens in that it stood above periodic inundation caused by sea and rivers. However, it could be flooded on high spring tides heightened by strong south-westerly gales blowing up the funnel-shaped opening of the Bristol Channel; hence substantial sea walls were essential. However, other than those walls only field-boundary ditches were necessary to carry rainfall either westwards towards the sea or eastwards towards the lower-lying inland peat moors. All in all, the clay belt displayed a normal conventional manorial economy, both in terms of land use and also of estate organisation.

It was in the inland peat moors, therefore, that reclamation occurred, usually taking the form of upgrading common grazing pasture in more favourable areas into meadow in order to produce an occasional hay crop. In a few cases, however, the land was drained so thoroughly that cultivation could be undertaken. Occasionally *rhynes* (open drainage ditches) were dug across the surface of the new ground, but more often walls were erected to keep out the regular and severe late autumn and winter floods. Everywhere, the clay-covered, often flood-prone, soils were preferred for upgrading because of their fertility, and the peat soils were left for common grazing or turbary.

The main area of reclamation was in the Parrett Valley in the vicinity of the junction of the rivers Tone and Parrett, and particularly in the Glastonbury manor of 'Sowy', roughly coincident with the present settlements of Westonzoyland, Chedzoy and Othery, which in the thirteenth century sported a flourishing conventional open-field system on the flood-free land of Sowy island. But there was lack of pasture land and it was in the marsh areas to the south of the 'island' and abutting the Parrett that efforts were made to reclaim land during the early thirteenth century, first by piecemeal temporary enclosures and later by attempting to exclude the flood waters of the Parrett and its tributaries by embanking the low-lying moors and surrounding them with high walls which also served as causeways linking the islands and peninsulas with Glastonbury. Such were Aller Wall beside the Parrett, Beer Wall linking Sowy with Glastonbury, Burrow and Southlake Walls around Southlake Manor, and Lake Wall blocking the northern edge of the Weston Level. At the same time as these walls were built, the

Figure 4.5: Reclamation in the Medieval Levels of Somerset

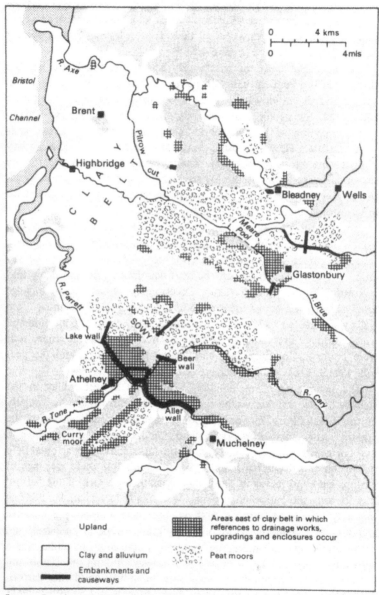

Source: Based on M. Williams, *The Draining of the Somerset Levels* (Cambridge University Press, Cambridge, 1970), by permission of the Cambridge University Press.

wandering and sinuous courses of the Parrett and Cary were straightened to improve the flow of water and reduce the length of wall that had to be built. Draining, therefore, proceeded in a number of literally water-tight compartments; inside the encircling walls, the moors were upgraded and surplus water let out into the rivers via sluices at low water.

Charters and rent rolls of the Glastonbury estates show clearly how the lowland was divided up into small blocks for which rents were paid by the tenants. Unlike the Fenland, reclamation was directed from above, and often came about because the Abbey wanted to reassert its control over the many illegal enclosures and 'purprestures' (encroachments on waste) made in the moors and to initiate some of its own reclamation for the obvious profit to be gained.

What happened to the Sowy estates was repeated in the moors around Glastonbury itself, where many small parcels of land were upgraded and reclaimed from the clay soils on the edges of the peat moors. Perhaps more important here were the major river diversions; flooding in the Levels demanded the rapid evacuation of water. It seems probable that the present river Brue is artificial and that the original river flowed northwards through the Panborough-Bleadney gap and filled the present course of the Axe. This is suggested by the presence of old meanders and defunct river channels that can be detected on air photographs, by sinuous freshwater deposits where no rivers run today, by the precise level (in feet) prepared during the reconnaisance of the M5 motorway routes across the clay belt, and by what one can see in the landscape itself. The westwards continuation of the Brue, via Meare Pool, probably came much later in the thirteenth century, and may have arisen as a part of the construction of the barge canal of the Pilrow Cut, excavated sometime before 1326, as a link between the outlying Glastonbury estates of Brent on the clay belt and Glastonbury Abbey. The further continuation of the Brue westwards to its present outfall at Highbridge came later on in the fifteenth century, and the new outlet certainly helped rid the Brue valley of its pent-up water behind the clay belt and so improved drainage as far back as Glastonbury.

Other Marshland Areas

The reclamation of the marsh was not confined, of course, to the Fens and the Levels, although these were the largest areas affected. In the south-east of England the priories of Bilingstone and Canterbury Cathedral were busily 'inning' land from the sea and estuaries in Romney Marsh, the adjoining Walland Marsh, the Isle of Thanet, and the

southern banks of the Thames at Sheppey. The obligation to maintain
sea walls appears in documents as early as 1100, and by the middle of
the thirteenth century a comprehensive *consuetudo et lex marisci* had
been built up, which became the model for other marshland areas. It
is difficult to know exactly, however, what was done and how much
marsh was reclaimed; we can only calculate that if Romney Marsh
contains about 17,300 acres between the Great Wall at Appledore and
the sea, then 7,144 acres were owned by Canterbury in the late thirteenth century, of which the Priory held 1,690 acres, all of which must
have been won from the sea.[27]

Reclamation was not everywhere an outstanding success, however.
For example, the thirteenth-century accounts of the prior's manor of
Ebony, which coincides more or less with the Isle of Oxney, reveal that
just over 14 per cent of the entire annual income was spent on embanking and draining, an expenditure which rose to as much as 60 per cent
during the years of 1287-90 when severe storms caused serious damage
to the sea walls. The same held true for other manors in the Marsh.
This constant maintenance has to be viewed against a background of
declining returns; the number of acres cultivated and the number of
stock pasturing fell continuously during the late fourteenth century.
One can only conclude that the soil was poor, an impression gained
from the frequent reference to marling and manuring, that the peat
soils kept shrinking and so needed extra embanking, and that the
recurrent inundations all took their toll. On the other hand, in other
manors such as Lydden, south of Sandwich, and Monkton in the Isle
of Thanet, coastal reclamation proved to be very profitable.[28]

Further along the coast at Pevensey, the Abbey of Battle was
engaged in ditching and walling the estuary that lay between the dry
ground of Barnhorne and Pevensey and so joining the salt marshes
that lay between the low clay and shingle 'islands' that accumulated
in the sheltered waters. The presence of salt mounds about two miles
inland at Norteye suggests that the previous coastline was further inland
in early medieval time.[29]

In the north, the Cistercian Abbey of Meaux was situated on an
'island' in the middle of the Hull Valley and almost totally surrounded
by marshes. Reclamation began in at least the first three decades of the
thirteenth century by digging major water courses to carry the surplus
water that drained off the surrounding uplands, such as Forthdyke,
which was 16 feet wide and 6 feet deep and was used for navigation
purposes initially, but soon developed drainage functions, as did all
navigation cuts. Along the Humber banks substantial walls were erected

to keep out tidal inundations. Further inland, Fountains Abbey held large tracts of land alongside the Ouse in the low-lying Vale of York, and the evidence is clear that large areas of land were ditched, banked and upgraded into meadows.[30]

The Marshland in the Later Middle Ages

Most of the evidence of reclamation in the marshland areas occurs before the plagues of the early thirteenth century, and there is very little information for later years. On the whole, one would expect a slowing down of reclamation as the pressure of population abated and the economic downswing which followed got under way. There was also a quickening of the process of commutation and leasehold, and a change from demesne exploitation to a rentier economy. Large estates battled for profits, and individual landholders probably let go some of their outlying clearings and precarious reclamations to concentrate on the more profitable ones of long standing.

The general impression is that during the late fourteenth and early fifteenth centuries little new land was reclaimed from the waste of the marsh. In the Fens the excavation of Morton Leam, a straight 12 miles long and 40 feet wide, out from Stanground near Peterborough to Guyhirne, to cut off the meanderings of the Nene and accelerate its flow by shortening its course, was outstanding, but an exception.[31] Rather it was a battle to find the manpower, the means and the material to do the everyday maintenance of the walls and ditches so laboriously built during the previous centuries – now threatened by deteriorating climatic conditions – and on which the sheer survival of the marshland areas depended. The co-operative effort of frontier colonisation in the Fens, and the estate engrossment and enlargement there and elsewhere by the ecclesiastical houses and large landholders, had to be superseded by an organisation capable of coping with the humdrum affair of routine maintenance: the settling of disputes which arose continuously over the non-performance of duties, the making of new works detrimental to the functioning of existing works, and the deliberate destruction of works which seemed to occur as a consequence of any sort of disagreement between neighbours and erstwhile friends. The 'ancient custom' of the marsh was creaking in the effort to achieve the hydrological control of these low-lying areas. Consequently, Royal Commissions of Sewers were issued, which empowered a jury to inquire about and examine the drainage works, which from about 1300 onwards hardened into a regular system of maintenance in nearly all marshland areas. Dugdale collected the evidence of the various commissions for every part of the country, and it is clear from these that local juries of landowners were established to enforce maintenance, the custom of which varied from place to place.[32] Sometimes the responsibility for maintenance was related to the amount of land held, sometimes to the length of frontage of work that lay adjoining the land held, and sometimes, as in Somerset, in relation to the number of beasts a freeholder pastured on the commons. The method of performance of the responsibility could be in day-labour, or more often, and increasingly as time wore on, in the form of a levy.

By 1472, the activities of the *ad hoc* commissioners were defined in an Act, but the function of the commissioners and the local juries changed little, the Act making *de jure* what had been *de facto* before. The laws of Romney Marsh regarding maintenance were applied widely in England, but little more was done than to maintain the works of rectification and repair.[33] It was a new dark age of endeavour compared to what had gone before.

Woodland Colonisation

The growing population pressed upon resources, not only in the marshland but also in the woodland. In nearly every part of the country the waste of scrub and high woodland on the edges of cultivation, both within the village territory and on the fringes of settlement — as, for example, in Devon, the Welsh borderland, Essex, the Wealden parts of Sussex and Kent, and even northern Warwickshire — was being reclaimed to produce more pasture, but mainly more arable land.

Yet, by the very nature of its growth habit, and as a consequence of the piecemeal clearing of previous centuries, the distribution of the remaining woodland was very patchy and mainly of small extent. Consequently, there was no concerted plan of colonisation, as in the Fens. Co-operative action was not possible in the woodland, nor indeed was it necessary; there was nothing comparable to the hydrological control and maintenance of works that characterised the lowland parts of the country. More often than not, assarting, or the clearing and grubbing up of the vegetation, was an individual, or at best a small-group, affair; it was slow and arduous, the product of 'a daily blinding sweat, blood at times, a back-breaking toil with axe and spade and saw'.[34]

Today the results of this effort can be identified in the landscape in a number of ways: by the evidence of documents which describe clearing; by the evidence of place-names which are indicative of cleared land or secondary settlement such as *-rydding, -leah, -feld, -stoc, -wood, -dene, -hurst, -holt, -stubbing*; and by the evidence of a mosaic of small, irregularly shaped fields, winding lanes and scattered settlement, which are the imprint of individual effort in the landscape. Sometimes all these strands of evidence coincide, and the story of the clearing of the woodland is abundantly clear.

The documentary evidence of clearing is copious. Outstanding in terms of area affected were the clearings of about 1,000 acres of the Bishop of Winchester's manor of Witney in Oxfordshire, and a similar amount in the manor of Wargrave in Berkshire, during the first half of the thirteenth century, followed by a further 660 acres and 680 acres, respectively, between 1256 and 1306. The arable land owned by the monks and tenants of Battle Abbey grew to 1,400 acres in the fifty years after its foundation, and a further 6 square miles of woodland were assarted at Rotherfield manor sometime between 1086 and 1346. In Laughton, Sussex, nearly 1,000 areas were added to the cultivated land between 1216 and 1325, and in the estates of Ramsey Abbey in Huntingdonshire, the heavy clay lands of the interfluvial uplands in the manors of Upwood and Warboys (Warbois) were cleared systematically, as were some 350 acres of assarts held by about thirty peasants for money rents in the Abbey's manor of Cranfield, Bedfordshire.[35]

These were large amounts of woodland cleared on large estates for which records of improvement were kept, but generally most of the clearing was done by free peasants, tenants with a licence from their lord to go ahead and open up the land. These small-scale piecemeal assarts in the forest were usually not recorded and if they were they can rarely be followed in detail, even for large areas of contiguous woodland, as in Arden in northern Warwickshire or the Sussex Weald or the lower hill slopes of Devon. Consequently, other types of evidence must be brought to bear in our analysis of woodland colonisation.

The way in which a variety of sources and approaches assists in the task of understanding woodland colonisation is provided by the example of the Forest of Arden in Warwickshire.[36] In his study of the clearing and settlement of Arden, Roberts concentrates on the 9,400 acres of the parish of Tanworth in the heart of the Forest. He locates the charters that record clearing, thereby reconstructing tentative maps of the various landscapes between about 1150 and 1350, supplemented by the evidence of court rolls, rent rolls and later surveys. The picture which emerges from this detailed study of one parish is of small freeholders, sponsored by the Earls of Warwick, colonising the waste, first near the open fields and then later in either the virgin or the cutover woodland. Thus, in the late twelfth century, Waleran, Earl of Warwick, granted Herbert, son of Dolfin 'all the land in Werdesworth which Dolphin held of the grantor', and this was typical of the early seigneurial grants of 2-15 acres which were rented for about two pence per acre. Later, the grants were as large as 60 acres and were located further away. In this way colonisation and clearing progressed northwards in waves from the early established nuclei of Anglo-Saxon villages, leading in time to the creation of a characteristic pattern of small cottages and single farmsteads, intermingled with small fields, the whole in a piecemeal and irregular order. Occasionally hamlets with names like 'Green', 'Ends' or 'Street' emerged at crossroads. In places, the more successful and prosperous farmers expanded their holdings to create small estates, like the Archers of Tanworth who by the mid-thirteenth century owned over one-third of the parish and created a sub-manor. Often such people built a substantial moated homestead, so much so that the moated farmstead is a distinctive feature of settlement in the once-wooded Arden landscape, as it is in other clay, late-colonised and once-forested areas, such as Essex and Norfolk.[37]

The characteristic features that emerge in the landscape of clearing in a single parish in Arden can be plotted for the whole of the county, and they tend to corroborate the documentary evidence. These features are plotted in Figure 4.6; first the extent of Domesday woodland, second the evidence of place-names such as *-ley*, which are indicative of clearing, and third the distribution of moated farmsteads.[38] All evidence shows the striking dichotomy between the north and the east, and the south and the west, between the late-cleared woodlands of Arden and the old-established manorial economy of the richer lands of the Felden. The coincidence of these distributions strengthens our belief in the evidence used to reconstruct the story of the colonisation of the woodland.

Perhaps more typical of our knowledge of woodland clearing is the fragmentary evidence which we have of the smaller piecemeal clearings in various parts of the country. For example, Brandon traces the clearing, charter by charter, in Framfield and Rotherfield manors in the Sussex Weald, which was all part of freeholder activity with the encouragement of landowners anxious to secure profits.[39] Here, as in Arden, the evidence of place-names is important with the proliferation of hamlets bearing the names ending in *-leah*, *-dene* and *-field*, suggesting late woodland colonisation. The Chiltern and Cotswold uplands were

Figure 4.6: The landscape of woodland clearing in Warwickshire: *a* the extent of Domesday woodland; *b* place-names indicative of clearing; *c* moated sites

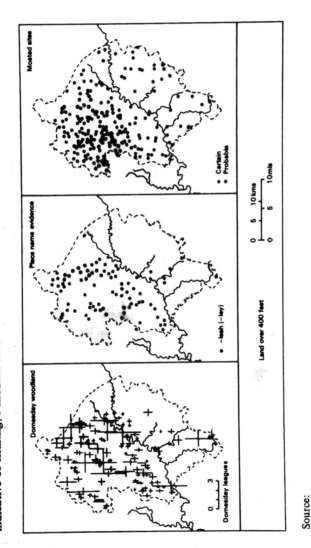

Source:
a. H.C. Darby and I.B. Terrett (eds.), *The Domesday Geography of Midland England*, 2nd edn (CUP, Cambridge, 1971), p. 296.
b. J.E.B. Glover et al., *The Place-names of Warwickshire* (CUP, Cambridge, 1936).
c. B.K. Roberts, 'Moated Sites', *Amateur Historian*, vol. 5 (1961–63), p. 40.

also well wooded in Domesday times, and the cartulary of Missenden Abbey and the Red Book of Worcester bear witness to the clearing that took place during the twelfth and thirteenth centuries. Peasant and landlord activity also explain the contemporary inroads into the wood and the extension of arable in the Blackmoor parishes of Stalbridge and Pulham in Dorset, while in the royal forest of Melchet the Wiltshire parish of Whiteparish was gradually extended.[40]

Not all woodland areas, however, were open to colonisation. In many parts of the country there was a conflict between landowners who wanted to retain their woodlands and freeholding peasants who wished either to expand their holdings or create new ones. The Statute of Merton (1236) recognised the lord's right to occupy and enclose the commons and waste, provided he left enough pasture for his free tenants, but 'enough' was difficult to define. Another source of conflict was that some monasteries were given extensive grazing rights at the expense of the villagers. But the main cause of conflict was preservation by the Crown of large areas under forest law for rough pastures, wood supplies and the chase. The areas affected are dealt with in detail in Chapter 3 above, but suffice it to say that the royal forests reached their greatest extent under the reign of Henry II (1154–89) to cover about one-third of the country, and that the Crown guarded its rights jealously and prosecuted those who trespassed, poached or assarted.[41] Nevertheless, by the early thirteenth century, the forests were beginning to appear like islands of undeveloped land in a sea of cultivation, and the temptation by agriculturalists wanting new land to encroach on the royal reserves was great, to say nothing of the inroads of the wood-getters and charcoal-burners. People offered the Crown money to be able to make assarts, and the Crown was also benefitting from the fines levied on illegal assarts made in the past. How much land was involved we do not know for sure; there are few detailed examples of assarting over a long period of time except for those of the forests of Cliffe and Rockingham, and Salcey and Whittlewood in Northamptonshire and Rutland which are shown in Table 4.1.

The Crown was also keen to get its hands on lump sums of ready money, just like any of the lesser lords. Consequently, during the thirteenth century large parts of the forests in Cornwall, Devon, Gloucestershire and Lincolnshire were opened to colonisation, and it is calculated that between 1250 and 1325 the overall forest area probably decreased by approximately one-third. For example, the whole of Devon and Cornwall had been disafforested in 1204 (except for Dartmoor

Table 4.1: Royal forests: assarting (in acres)

Forest	Date	Old assarts	New assarts	Total assarts
Cliffe and	1253	–	–	784
Rockingham	1255	–	–	780
	1272	363	–	363
	1339	753	487	1,240
	1346	634	554	1,188
	1355	697	486	1,183
Rutland	1344	714	236	950
	1375	283	1,126	1,409
Salcey and	1249	378	–	378
Whittlewood	1252	–	–	393
	1255	–	–	238
	1345	–	–	1,128
	1348	–	460	460

Source: C.R. Young, *The Royal Forests of Medieval England* (Leicester University Press, Leicester, 1979), p. 121.

and Exmoor), the men of Devon and Cornwall paying 5,000 and 2,200 marks respectively to do this. All these moves were a measure of the land shortage and consequent disafforestation was a stimulus to settlement and colonisation.[42]

Upland Colonisation

There were of course, other areas of colonisation, particularly in the highland zone of England which was still largely unsettled by 1086. Many areas above about 800 feet were devoid of Domesday settlement which, while not conclusive, is certainly indicative of a thinly populated area. In the north, the whole stretch of the Pennines, and the North York Moors and Rossendale are blanks on the Domesday map; and in the south-west, Dartmoor, Exmoor and to a lesser extent Bodmin Moor and the Quantocks are similarly blank, although the milder climate here enabled settlement to move further up the hill slopes; for example, there are some farms at 900 feet on the wet, west side of Dartmoor, but up to 1,200 feet on the drier eastern side.

Everywhere, the growth of population and the pressure to reclaim more land caused settlement to creep up the lower slopes, and particularly up the sheltered valleys that penetrated the upland areas, often through a zone of trees before reaching the open moorland. Permanent settlement for arable farming did not necessarily become established immediately, but sometimes followed many years after initial penetration by temporary settlements or shielings. The obstacles to farming in the uplands were different to those in the marshes and the woodland: standing water and trees were replaced by steep slopes, infertile soils, and rock and boulder-strewn surfaces. Once cleared of stones, the turfs might be cut, piled up and burnt, the ashes dug into the soil and an occasional grain crop of oats or rye taken off.[43] Sometimes the reclamation would be permanent, but often it was abandoned, reverted to rough pasture, and a new intake was made elsewhere.

The volumes of the English Place-Name Society, compiled for upland counties such as Cornwall, Devon, Westmorland and Northumberland, are packed with topographical names evoking the nature of the difficult terrain and distinguished by a late date of first recording. For example, in Devon Hoskins suggests that of roughly 15,000 farms, place-name evidence indicates that four-fifths were created after 1086, and that 'the great majority of these farms had come into existence by 1350 in the great colonization movement that had begun in the 12th century, and reached its peak in the 13th'.[44]

Typical of these farms was Cholwich (the coldest farm) in the parish of Cornwood on the south-western edge of Dartmoor, first recorded in about 1220-5 (Plate 4). It is a single compact farm on the moorland edge, representative of the individual outlying farms created by settlers as they took into their territory tracts of land which had probably already been used occasionally and casually as temporary outfields of the nucleated settlement of Cornwood that nestled in the sheltered lowlands below. The 200 acres of *Cholleswyht* were granted by the Lord of Cornwood to Benedict, son of Edric Siward, a small peasant freeholder. The great granite boulders of the farm house, and the foundations of the hedgerows that line the lanes must have been hauled off the rough ground to create the little fields.[45]

Further up the slopes and away from the sheltered valleys the high open moorland was used for summer grazing. Cattle thrived here, as they did on the lowland marshes, and the development of vaccaries around the Pennines was a notable feature. Donkin has identified fifteen cattle stations in Wyesdale and nine in Nidderdale belonging to Fountains Abbey, and several in Wensleydale belonging to Jervaulx, and similar developments by ecclesiastical houses and individual landholders characterised the Forest of Rossendale. The Cistercians also pastured large flocks of sheep in many upland areas, the wool being a major factor in the wealth of these religious houses, although sheep were more numerous on the lower hills like the Cotswolds, and the

Plate 4: The journey to the margin of cultivation: Cholwich Farm, Dartmoor, Devon, a farm carved out of the edge of the moorland in the early thirteenth century

Lincolnshire and Yorkshire Wolds. In many ways, the vaccary was a kind of pastoral out-station and was established for the colonisation of the upland areas, in much the same way as the grange was established for the colonisation of the lowland areas.[46]

In time, the level of stocking in the upland pastures increased, which had two effects: some summer camps became permanent, but more importantly the settlements surrounding the moors began to dispute the rights of neighbours to depasture their sheep or cattle as, indeed, was already happening in the lowland marsh areas. Thus, many high moorland areas were marked by stone walls in order to delineate property rights. For example, the rough pastures of the Kilnsey and Malham Moors above Wharfedale, which were in dispute by the tenants of Fountains, Sallay and Bolton Abbeys at various times during the late thirteenth century, were ultimately divided by miles of demarcating stone walls.[47]

It is difficult to generalise about what happened in these upland areas as each valley and interfluve presented a different picture of colonisation. A bewildering variety of patterns of colonisation are present in the north of England. In the upland mass of millstone grit known as East Moor in Derbyshire, the patterns of parish boundaries show how the girdle of older settlements shared all the upland area. Cattle farmers and piecemeal peasant enclosures nibbled at the edges of the moors, something which can be gleaned from the distribution of references to reclamation and assarting, and the distribution of the place-names containing elements of *-ridding* (land rid of trees), *-stubbing* (land with stumps removed) and *-intake* (land enclosed from the waste), all of which bespoke of individual enclosures of the waste on a piecemeal basis.[48]

In the uplands of Ryedale on the North York Moors, the evidence suggests a more comprehensive and planned exploitation than in East Moor (Figure 4.7). The parish boundaries again reflected the interdependence between upland pastures and lowland arable and meadow, but there were other features. The vills devastated during the 'scorched earth' policy of William I as he put down the opposition of the north during 1069 and 1070 were recolonised by lay and ecclesiastical owners, who comprised two distinct components of land-ownership. By the end of the thirteenth century the majority of the devastated vills were recolonised, and some new vills established. Of the 37 original Domesday vills, 14 were colonised by lay owners, eleven by ecclesiastical owners, five had mixed ownership, three were never restored, and four were converted into granges by their ecclesiastical owners. The lay owners subsequently established about thirty frontier farms and nearly half-a-dozen hamlets on or near the 800-feet contour on the moorland edge.

In contrast to this piecemeal activity of the freeholding peasantry, the large monastic houses went about exploiting the uplands in a much more methodical way. The Cistercian houses of Rievaulx, Rosedale and Keldholme acquired large block grants of land in the more favourable lowland valleys of Rosedale, Ferndale and Bilsdale respectively, and on the north-east portion of Ryedale the Benedictines of St Mary's, York had a large block of land. The Rievaulx block was exploited rationally with the establishment of eight granges and seven other smaller holdings, while the Gilbertine house of Malton (outside the area) had five granges in the south.[49]

The grange was a distinctive feature of the colonisation of the uplands of the north of England; however, its distribution was far more widespread than that, and it was to be found everywhere where the colonisation of the waste was in progress during the early Middle Ages,

Figure 4.7: Medieval colonisation in north Ryedale, North York Moors:
a. Elements of lay settlement; *b*. elements of monastic settlement

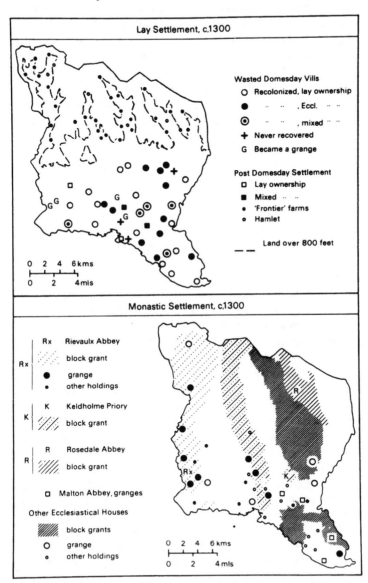

Source: Based on R.I. Hodgson, 'Medieval colonization in northern Ryedale, Yorkshire', *Geographical Journal*, vol. 135 (1969).

such as on the Fen edge, and as such commands our attention as a facet of the 'waste'. The grange consisted of a collection of farm buildings on a compact holding and it served as a sub-centre for the systematic colonisation of unused land by the mother settlement.[50] It was probably the invention of the Cistercian Order, but it soon spread to other religious houses. Often the farm was large (commonly 400–600 acres, but often larger) managed by lay brethren of the religious order and worked by a dependent peasantry. It was usually associated with grain farming, hence the name grange, taken from 'granary', the large storage barn of the produce of the farm. The grange represented a new form of territorial organisation for the aggrandisement of property and the pursuit of profits.

In all, some 350 Cistercian granges were created in England, over two-thirds of which were situated north of a line linking the Severn with the Wash. Clearly it was more difficult to assemble large consolidated estates in already well-settled areas, but much easier where there was abandoned 'wasted' ground or virgin waste ground that could be assarted, as in the north of England. For example, of the 120 granges in Yorkshire at least three-quarters were established before 1200, over 40 per cent of them on the 'wasted' land from the devastation of 1069–70. In addition, some weakened townships were forcibly taken over by the religious houses in order to consolidate their estates, but the rest were created from the waste. Consequently, the number of granges per house tended to be greater in the north than in the south of the country. In Yorkshire Rievaulx had thirteen granges, Jervaulx four, Byland thirteen, Kirkstall eight, and Salley ten, which were all founded in the later twelfth century in the valleys which penetrate the Yorkshire uplands, on the very edges of settlement around the Pennines and the North York Moors. On the one side, there was the vast expanse of the open moorlands, the lower sheltered parts enclosed and reclaimed, the upper parts converted into pastoral holdings. On the other side were the settled and devastated arable lands that were taken over, or even exchanged in order to consolidate the holdings. Fountains with its twenty-six granges was the largest of them all, and created much arable land by reclaiming the rich and varied soils in that portion of the Vale of York where it narrows between the houses of Fountains, Byland and Jervaulx. In the low-lying and floodable Hull river valley, Meaux with twenty-two granges engaged in much coastal and marsh reclamation, but it also had granges on the upland of the Wolds for sheep grazing.[51]

Epilogue: The Set-back

Enough has already been written to indicate that the general expansion of settlement into the waste was beginning to slow down even before the advent of the Black Death, and that by the early fourteenth century it was all but over. Over-population had pushed the pioneering settlers into difficult areas, and yields fell and holdings were subdivided into units that were probably below subsistence level. Declining productivity must have been felt most acutely on the frontiers of cultivation, that is on the marginal land on the extremes of settlement, whether in the marshlands, the woodlands or the uplands, although difficult land even in the midst of the well-settled areas was also vulnerable as the *Nonarum Inquisitiones* of 1341-2 show. Even in a favourable and varied environment like the Lincolnshire Fenland it is noticeable that although population continued to grow until the Black Death, there were no new reclamation schemes after the mid-thirteenth century.[52]

The problems of subdivision and declining yields are dealt with more extensively in other parts of this book, but there are two trends which were becoming evident before the plague that had a bearing on the waste lands. First, everywhere arable had expanded at the expense of pastoral land, and it is argued that there was a decline in livestock numbers.[53] This could have had two detrimental consequences: first, there could have been a protein deficiency in diets leading to a less healthy and virile population, and there would have been less manure available to bolster up the yields which appear to have been declining anyway, but most markedly on the newly colonised lands.[54]

Second, there may well have been a long-term deterioration of the climate. The relatively warm period of the eleventh and twelfth centuries was coming to an end by the closing decades of the thirteenth. The weather was generally colder and wetter, or at least of greater extremes, which affected harvests and caused murrain in stock and would have drastically affected settlement on the climatically sensitive margins of cultivation.[55] There is well-documented evidence of the extension of cultivation in the upland moors of the Lammermuir Hills during the twelfth century and its abandonment during the fourteenth and fifteenth, and it is not difficult to imagine how critical variations of only a few degrees could alter the precarious foothold of the pioneer agriculturalists in other localities further south in upland England, for example in the North York Moors and Dartmoor.[56] Along the coast the increased incidence of storms and inundations in Sussex, Kent, the Fens and the Broads leading in places to permanent changes in the relative levels of land and sea and the loss of newly reclaimed land underlined the vulnerability of these marginal lands.[57]

Not related to these areas of erstwhile waste is the evidence of land abandonment in the *Nonarum Inquisitiones* of 1341-2 due to soil exhaustion, and consistently unfavourable climate. Land reverted to grass from grain as the population declined and conditions were more variable.[58] The disappearance of over 1,200 villages from the English countryside from the late fifteenth century onwards was the tangible and quantifiable culmination of these changes.[59]

Thus, by the middle of the fourteenth century, the Black Death and the plagues which followed only confirmed and compounded the problems which were already becoming evident as the centuries-long effort to cultivate the waste drew to its close. In these circumstances of misery, fear and desolation it was not surprising that men's minds did not turn to making new lands from the waste when so much lay idle and deteriorating around them. Other and more profound problems concerning the structure of society were of more immediate concern.

To look at the recession as an end to the changes which had characterised the previous two centuries, however, is to miss the point that it was also a beginning that marked the start of an equally long drawn-out and painful transition that culminated in the ebullience and sparkle of the early Tudor times. The forces at work during these later Middle Ages were almost as complex and varied, and are as barely understood, as those during the early Middle Ages. Population levels remained low and pressure on the land was minimal. Little new land was brought into production, rather a massive reorganisation occurred as demesne farming declined and peasant holdings increased, as feudal powers diminished and peasants became more mobile and assertive, and as the rural settlement pattern was reorganised (usually deserted) in response to demographic decline, economic influences and land changes through enclosure. Nearly every urban centre, however, seemed to acquire new influence and importance as many peasants migrated to the towns, and textile and extractive industries flourished. The exact chronology and causality of these affairs is open to lively debate, and there is doubt if the latter Middle Ages overall was a period of growth, decline or stagnation, but the general picture is one of labour scarcity and land surplus and consequently little, if any, expansion.[60]

And yet, prosperity did not remain static. Schofield has shown that the south-west counties of Cornwall, Devon and Somerset, and the south-east counties of Herefordshire, Essex, Kent, Suffolk and Middlesex increased their wealth by over 400 times between 1334 and 1515.[61]

The work of Darby *et al.* (Figure 4.2) confirms the general distribution but pin-points the areas of massive growth: the south-west peninsula (mining and some reclamation) together with adjacent cloth-manufacturing in parts of Somerset and Wiltshire; the cloth-manufacturing areas of Essex and Suffolk; part of the West Riding cloth-manufacturing area; London and its surrounding areas, and the peat lowlands of the southern Fenland. Only in the first and last areas was there agricultural expansion in the way it was experienced before, but even so there is little detailed evidence forthcoming.[62] The next time the English farmer journeyed to the margin in a way that was comparable to the early Middle Ages was on the eve of the Industrial Revolution, when from about 1750 and for the next hundred years the margins of cultivation were extended to their uttermost limits.[63]

Notes

1. H.C. Darby, *Domesday England*, (Cambridge University Press, Cambridge, 1977), pp. 36–45.

2. M.W. Beresford and J.K. St Joseph, *Medieval England: An Aerial Survey*, (Cambridge University Press, Cambridge, 1958; rev. edn, 1979), pp. 91–3.

3. J.C. Russell, *British Medieval Population* (New Mexico University Press, Alberqueque, 1940), pp. 34–54; M.M. Postan, 'Medieval Agrarian Society at its Prime: England' in M.M. Postan (ed.), *Cambridge Modern History of Europe*, vol. 1, 2nd edn. (Cambridge University Press, Cambridge, 1966), pp. 560–5; J.M.W. Bean, 'Plague, Population, and Economic Decline in England in the later Middle Ages', *Economic History Review*, 2nd ser., vol. XV (1963), pp. 422–37; B.F. Harvey, 'The Population Trend in England between 1300 and 1348', *Royal Historical Society Transactions*, 5th ser., vol. 16 (1966), pp. 23–42; and J. Hatcher, *Plague Population, and the English Economy, 1348-1530*, (Macmillan, London, 1977), pp. 21–5, 69–72. For a summary of the evidence on the pre-Black Death population see J.Z. Titow, *English Rural Society, 1200-1350* (George Allen and Unwin, London, 1969), pp. 66–73.

4. G. Duby, *Rural Economy and Country Life in the Medieval West* (Edward Arnold, London, 1968), pp. 119 ff.

5. Postan, 'Medieval Agrarian Society at its Prime: England', pp. 552–9; A.H.R. Baker, 'Evidence in the *Nonarum Inquisitiones* of Contracting Arable Lands in England during the early fourteenth century', *Economic History Review*, 2nd ser., vol. XIX (1966), pp. 518–32.

6. H.E. Hallam, *Settlement and Society: A Study of the Early Agrarian History of South Lincolnshire* (Cambridge University Press, Cambridge, 1965), pp. 200–22; 'Some thirteenth century Censuses', *Economic History Review*, 2nd ser., vol. X (1958), pp. 340–61; and 'Population Density in the Medieval Fenland', *Economic History Review*, 2nd ser., vol. XIV (1961), pp. 71–81; and R.A. Donkin, 'Changes in the early Middle Ages' in H.C. Darby (ed.), *A New Historical Geography of England before 1600* (Cambridge University Press, Cambridge, 1973), pp. 75–81.

7, J.B. Harley, 'Population Trends and Agricultural Developments from the Warwickshire Hundred Rolls of 1279', *Economic History Review*, 2nd ser., vol.

XI (1958), pp. 8–18; and 'Settlement Geography of early medieval Warwickshire', *Transactions, Institute of British Geographers*, vol. 34 (1964), pp. 115–30.

8. E. Miller and J. Hatcher, *Medieval England: Rural Society and Economic Change, 1086–1348* (Longman, London, 1978), pp. 31–3.

9. H.E. Hallam, 'The Postan Thesis', *Historical Studies*, vol. 15 (1972), pp. 203–22. Hallam supports an explanation that technological advance needed population pressure. He asks (p. 222), 'Are we rich because we are numerous?'

10. W.G. Hoskins, *The Making of the English Landscape* (Hodder and Stoughton, London, 1957), p. 76.

11. E.J. Buckatzsch, 'The Geographical Distribution of Wealth in England, 1086–1843', *Economic History Review*, 2nd ser., vol. III (1950). pp. 180–202; and R.S. Schofield, 'The Geographical Distribution of Wealth in England, 1334–1649', *Economic History Review*, 2nd ser., vol. XVIII (1965), pp. 483–510.

12. H.C. Darby, R.E. Glasscock, J. Sheail and G.R. Versey, 'The Changing Geographical Distribution of Wealth in England: 1086–1334–1535', *Journal of Historical Geography*, vol. 5 (1979), pp. 247–62.

13. Ibid., p. 248.

14. Ibid; and Michael Stanley, 'The Geographical Distribution of Wealth in medieval England', *Journal of Historical Geography*, vol. 6 (1980), pp. 315–20.

15. H.E. Hallam, 'The Postan Thesis', pp. 203–22.

16. For a thorough discussion on fishing, fowling, peat digging and other traditional occupations in the marsh, see H.C. Darby, *The Medieval Fenland* (Cambridge University Press, Cambridge, 1940), pp. 21–42, 82–5; and M. Williams, *The Draining of the Somerset Levels* (Cambridge University Press, Cambridge, 1970), pp. 26–32.

17. J.M. Lambert, J.N. Jennings, C.T. Smith, Charles Green and J.N. Hutchinson, *The Making of the Broads* (Royal Geographical Society Research Series, no. 3, London 1960).

18. Darby, *The Medieval Fenland*, pp. 67–83 and 86–92; Williams, *The Draining of the Somerset Levels*, pp. 32–8.

19. Darby *et al*. 'The Changing Geographical Distribution of Wealth in England: 1086–1334–1535', pp. 249–52.

20. Hallam, *Settlement and Society*.

21. See E. Miller, *The Abbey of Ely* (CUP, Cambridge, 1951), pp. 95–8, for evidence of marsh reclamation in Norfolk.

22. The following account of reclamation in the Fenland is based on Hallam, *Settlement and Society*, pp. 3–118, except where otherwise stated.

23. H.E. Hallam, 'Goll Grange, a Grange of Spalding Priory', *Lincolnshire Architectural and Archaeological Society's Reports and Papers*, vol. XXXXVII (1953) pp. 1–18.

24. Hallam, *Settlement and Society*, p. 103.

25. Ibid., pp. 174–96.

26. The following account of reclamation in the Somerset Levels is based on Williams, *The Draining of the Somerset Levels*, pp. 40–74.

27. R.A.L. Smith, *Canterbury Cathedral Priory* (CUP, Cambridge, 1943), pp. 166–89; N. Neilson, *The Cartulary and Terrier of the Priory of Bilsington, Kent* (British Academy, London, 1928), pp. 41–56; N. Harvey, 'The Inning and Winning of the Romney Marshes', *Agriculture*, vol. 62 (1952), pp. 334–8.

28. Smith, *Canterbury Cathedral Priory*, pp. 178, 184–6.

29. L.F. Salzmann, 'The Inning of Pevensey Levels', *Sussex Archaeological Collections*, vol. 53 (1910), pp. 32–60; P. Brandon, *The Sussex Landscape* (Hodder and Stoughton, London, 1974), pp. 111–18.

30. J.A. Sheppard, *The Draining of the Hull Valley* (East Yorkshire Local History Series, York, 1958); S.G.E. Lythe, 'The Organization of Drainage and Embankment in medieval Holderness', *Yorkshire Archaeological Journal*, vol.

34 (1939), pp. 282–95; R.A. Donkin, 'The Marshland Holdings of the English Cistercians before c. 1350', *Citeaux in de Nederlanden*, IX (1958), pp. 262–75; *The Cistercians: Studies in the Geography of Medieval England and Wales*, (Pontifical Institute of Medieval Studies, Toronto, 1978), pp. 117–18. For other marshland areas not considered in detail here see J. Thirsk, 'The Isle of Axholme before Vermuyden', *Agricultural History Review*, vol. I (1953), p. 16; D.B. Hardman, 'The Reclamation and Agricultural Development of North Cheshire and South Lancashire Mossland Areas', unpublished MA thesis, University of Manchester, 1961; and B.E. Cracknell, *Canvey Island: The History of a Marshland Community* (University of Leicester Occasional Paper, 12, Leicester, 1959).

31. Darby, *The Medieval Fenland*, pp. 167–8.

32. W. Dugdale, *History of Imbanking and Drayning* (London, 1662).

33. H.G. Richardson, 'The early History of the Court of Sewers', *English Historical Review*, vol. XXXIV (1919), pp. 385–93.

34. W.G. Hoskins, 'The English Landscape' in A.L. Poole (ed.), *Medieval England*, vol. I (OUP, Oxford, 1958), p. 10.

35. J.Z. Titow, 'Some Differences between Manors and their Effects on the Conditions of the Peasantry in the thirteenth century', *Agricultural History Review*, vol. X (1962), pp. 4–8; Eleanore Searle, *Lordship and Community: Battle Abbey and its Banlieu* (Toronto University Press, Toronto, 1974), pp. 58–60; J.S. Moore, *Laughton: A Study in the Evolution of the Wealden Landscape* (University of Leicester Occasional Paper, 19, Leicester, 1965), pp. 41–2; J.A. Raftis, *The Estates of Ramsey Abbey, A Study of Economic Growth and Organization*, (Pontifical Institute of Medieval Studies, Toronto, 1957), pp. 71–4, and footnote 30 on pp. 4–5.

36. B.K. Roberts, 'A Study of medieval Colonization in the Forest of Arden, Warwickshire', *Agricultural History Review*, vol. XVI (1968), pp. 101–13.

37. B.K. Roberts, 'Moated Sites', *Amateur Historian*, vol. 5 (1961–3), pp. 34–8, 40; F.A. Arberg, *Medieval and Moated Sites* (Research Report No. 17, Council for British Archaeology, London, 1978), pp. 47–9.

38. H.C. Darby and I.B. Terrett (eds.), *The Domesday Geography of Midland England*, 2nd edn. (Cambridge University Press, Cambridge, 1971), p. 296; J.E.B. Glover, *et al.*, *The Place-names of Warwickshire* (Cambridge University Press, Cambridge, 1936); and Roberts, 'Moated Sites', p. 40.

39. R.F. Brandon, 'Medieval Clearances in the East Sussex Weald', *Transactions, Institute of British Geographers*, vol. 48 (1968), pp. 135–53.

40. E.C. Vollans, 'The Evolution of Farm Lands in the Central Chilterns in the twelfth and thirteenth centuries', *Transactions, Institute of British Geographers*, vol. 26 (1959), pp. 220–2; Hoskins, *The Making of the English Landscape*, pp. 69–71, 73–5; C.C. Taylor, 'The Pattern of Medieval Settlement in the Forest of Blackmoor', *Proc. Dorset Nat. Hist. and Arch. Soc.*, vol. 87 (1965), pp. 252–4; and 'Whiteparish: A Study in the Development of a Forest-edge Parish', *Wiltshire Arch. and Nat. Hist. Mag.*, vol. 62 (1967), pp. 79–102; E.M. Yates, 'Dark Age and Medieval Settlement on the Edge of the Wastes and the Forests', *Field Studies*, vol. 2 (1965), pp. 133–53 has some excellent maps of encroachments on the edges of Needwood Forest in south-east Staffordshire, Dartmoor, and Morfe Forest, east Shropshire.

41. For the extent of the forests see M.L. Bazely, 'The Extent of the English Forest in the thirteenth century', *Transactions, Royal Historical Society* 4th ser., vol. 4 (1921), pp. 140–72; and N. Neilson, 'The Forests' in J.F. Willard and W.A. Morris (eds.), *The English Government at Work, 1327–1336*, vol. I (Medieval Academy of America, Cambridge, Mass., 1940), pp. 394–448.

42. C.R. Young, *The Royal Forests of Medieval England* (Leicester University Press, Leicester, 1979), pp. 116–24.

43. J. Hatcher, *Rural Economy and Society in the Duchy of Cornwall,*

1300-1500 (CUP, Cambridge, 1970), pp. 82-5.

44. W.G. Hoskins, *Devon,* (Collins, London, 1954), p. 70.

45. W.G. Hoskins and H.P.R. Finberg, *Devonshire Studies* (Cape, London, 1952), pp. 78-93; and Beresford and St Joseph, *Medieval England*, pp. 92-3.

46. R.A. Donkin, 'Cattle on the Estates of medieval Cistercian Monasteries in England and Wales', *Economic History Review*, 2nd ser., vol. XV (1962), pp. 31-53; *The Cistercians: Studies in the Geography of Medieval England and Wales*, pp. 68-85; and G.H. Tupling, *The Economic History of Rossendale* (Chetham Society, new ser., no. 86, Manchester, 1927), pp. 17-27.

47. A. Raistrick, *The Story of the Pennine Walls* (Clapham, 1946), chs. 6 and 7.

48. S.R. Eyre, 'The Upward Limit of Enclosure on the East Moor of North Derbyshire', *Transactions, Institute of British Geographers*, vol. 23 (1957), pp. 61-74.

49. R.I. Hodgson, 'Medieval Colonization in northern Ryedale, Yorkshire', *Geographical Journal*, vol. 135 (1969), pp. 44-54.

50. The work of R.A. Donkin is crucial to our understanding of the grange. See Donkin, *The Cistercians*, pp. 37-67; 'The Cistercian Grange in England in the 12th and 13th Centuries with special reference to Yorkshire', *Studia Monastica*, vol. 4 (1964), pp. 75-144. Other useful sources are T.A.M. Bishop, 'The Monastic Grange in Yorkshire', 4 parts, *Yorkshire Archaeological Journal*, vol. 37 (1948-51), pp. 474-91, and vol. 38 (1952-5) pp. 44-70, 215-40; 280-309; and B. Waites, 'The Monastic Grange as a Factor in the Settlement of north-east Yorkshire', *Yorkshire Archaeological Journal*, vol. 40 (1959-60), pp. 627-56.

51. For the maps of the Meaux grange estates see C. Platt, *The Monastic Grange in Medieval England* (Macmillan, London, 1969), pp. 49-75.

52. H.E. Hallam, 'Some thirteenth century Censuses', *Economic History Review* 2nd ser., vol. XI (1958), pp. 340-61.

53. M.M. Postan, 'Village Livestock in the thirteenth century', *Economic History Review*, 2nd ser., vol. XV (1962), pp. 219-49.

54. R.H. Britnell, 'Agricultural Technology and the Margin of Cultivation in the fourteenth century', *Economic History Review*, 2nd ser., vol. XXX (1977), pp. 53-66; J.Z. Titow, *Winchester Yields: A Study in Medieval Agricultural Productivity* (CUP, Cambridge, 1972), pp. 20-4, 32-3.

55. H.H. Lamb, *The Changing Climate, Selected Papers* (Methuen, London, 1966), chs. 1 and 7; see also J.Z. Titow, 'Evidence of Weather in the Account Rolls of the Bishopric of Winchester, 1209-1350', *Economic History Review*, 2nd ser., vol. XII (1959), pp. 360-407. It seems that 1315 and 1316 were exceptionally wet; there were harvest failures and disease was rife in cattle and sheep. I. Kershaw, 'The Great Famine, and Agrarian Crisis in England, 1315-1322', *Past and Present*, vol. 59 (1973), pp. 3-18.

56. M.L. Parry, *Climatic Change, Agriculture and Settlement* (Dawson, Folkestone, 1978) pp. 113-17, 122-5; M.W. Beresford, 'Medieval Agriculture' in A. Raistrick (ed.), *North York Moors* (National Parks Guide, London, 1969), p. 55; and Kershaw, 'The Great Famine and Agrarian Crisis in England, 1315-1322'; and G. Beresford, 'Three Deserted Medieval Settlements on Dartmoor', *Medieval Archaeology*, vol. 23 (1979), pp. 98-155.

57. A.R.H. Baker, 'Some Evidence of a Reduction in the Acreage of Cultivated Land in Sussex during the early fourteenth century', *Sussex Archaeological Collections*, vol. 104 (1966), pp. 1-5; Smith, *Canterbury Cathedral Priory*, pp. 150-3; Hallam, *Settlement and Society*, pp. 119-36; and Lambert *et al.*, *The Making of the Broads*.

58. Baker, 'Evidence in the *Nonarum Inquisitiones* of Contracting Arable Lands in England during the early fourteenth century', pp. 518-32.

59. M.W. Beresford, *The Lost Villages of England* (Lutterworth Press, London,

1954); and M.W. Beresford and J.G. Hurst, (eds.), *Deserted Medieval Village Studies*, (Lutterworth Press, London, 1971), pp. 182–212.

60. For a review of the period see A.H.R. Baker, 'Changes in the later Middle Ages' in H.C. Darby (ed.), *A New Historical Geography of England before 1600*, pp. 186–247; R.A. Butlin, 'The later Middle Ages' in R.A. Dodgshon and R.A. Butlin (eds.), *An Historical Geography of England and Wales* (Academic Press, London, 1978), pp. 119–50; and A.R. Bridbury, *Economic Growth in England in the later Middle Ages* (Harvester Press, Hassocks, Sussex, 1972).

61. Schofield, 'The Geographical Distribution of Wealth in England, 1334–1649', pp. 504–9, and map in Butlin, 'The later Middle Ages', p. 144.

62. Darby *et al.*, 'The Changing Geographical Distribution of Wealth in England: 1086–1334–1535', pp. 256–62 for a detailed discussion.

63. M. Williams, 'The Enclosure and Reclamation of Waste Land in England and Wales in the eighteenth and nineteenth centuries', *Transactions, Institute of British Geographers*, vol. 51 (1970), pp. 55–69.

5 CASTLES, FORTIFIED HOUSES, MOATED HOMESTEADS AND MONASTIC SETTLEMENTS

Leonard Cantor

In a predominantly rural society, organised on a hierarchical basis, the most prominent and permanent buildings outside the towns and villages were the castles, fortified houses, moated homesteads and monastic settlements — the residences of the Crown, the nobility, the great ecclesiastical lords and other less powerful landowners. Built in substantial numbers throughout the country, of timber, stone and, towards the latter part of the Middle Ages, of brick, they represented important often imposing elements in the medieval landscape. In treating each group of buildings separately, we must first attempt to define them, by no means an easy task given the inevitable degree of overlap and the confusing and sometimes conflicting use of terminology that is an all too frequent feature of English medieval history. However, as a valid generalisation, it can be taken that castles and fortified houses form a cognate group in which fortification is a major, if not the only, element in their construction and siting. Moated homesteads, on the other hand, had a primarily agricultural function in which the moat provided a degree of protection. Finally, monastic settlements owe their origin to the proselytising zeal of the monastic orders, though they too were not infrequently, in lawless times and in remote parts of the country, grateful for the degree of protection which such features as moats might afford.

Castles and Fortified Houses

For the purpose of this chapter, therefore, it is convenient to deal with castles and fortified houses together and we may define them as the private fortresses and residences of a lord, whether king, baron or great ecclesiastic. In doing so, we are drawing heavily upon the definitive work on English castles of R. Allen Brown, and especially his *English Castles*. As we shall see, this definition embraces three main types of building: the earth and timber castles, prevalent during the century following the Norman Conquest; the stone castles which began to become common after the middle of the twelfth century but which

126

were occasionally found earlier; and the fortified manor houses which were particularly prevalent during the late thirteenth and fourteenth centuries. Because all traces of the timber structures which made up the first category have long since vanished, it is really only the second group and, to a lesser extent, the third which are popularly recognised as castles.

Fortified towns have been a common and well-established military practice throughout history and, as in the case of the Anglo-Saxon *burh*, were communal in nature.[1] Castles and fortified houses, on the other hand, belonged uniquely to the Middle Ages and were distinguished by their private character. This in turn distinguished them from their successors, the purely military fortress of modern times. Thus, they performed two major functions: that of the fortress and that of the private residence, respectively military and domestic.[2] By and large, the erection and maintenance of castles and fortified houses was the prerogative of the Crown and the more powerful nobles. In order to raise a castle, the latter had to obtain a licence from the Crown which, for reasons of security in the years following the Conquest, encouraged the Norman barons to build them to hold down the territory which William had distributed among them. However, the owners of castles, whether royal or baronial, did not live permanently or regularly in them. Indeed, for most of the Middle Ages, they were so ceaselessly on the move with their retinues, visiting their estates, hunting, or campaigning, that it would be difficult to say that they lived anywhere at all.[3]

We are not here concerned with the architecture or military significance of English castles, about which there is a considerable and growing literature,[4] but with their contribution to the medieval landscape and economy, and their spatial distribution. The castle was of French origin and was built on the two basic principles of a fortified enclosure surrounding a strong tower. The enclosure, or 'ring-work' as the archaeologists would have it, commonly consisted of a ditch and bank topped by a timber palisade or stone wall, and the tower, keep or 'donjon', was usually of wood at first, subsequently of stone and, in the later Middle Ages, occasionally of brick. The castle was introduced into England by the Normans as a means of establishing a firm hold upon the conquered countryside. Consequently, the Conquest was followed by an enormous amount of castle-building, primarily of the *motte-and-bailey* type. The motte, or mound, consisted of earth or rubble, surrounded by a ditch and surmounted by a wooden tower. This looked down over a bailey, or courtyard, surrounded by a palisade

Figure 5.1: Castles in 1086

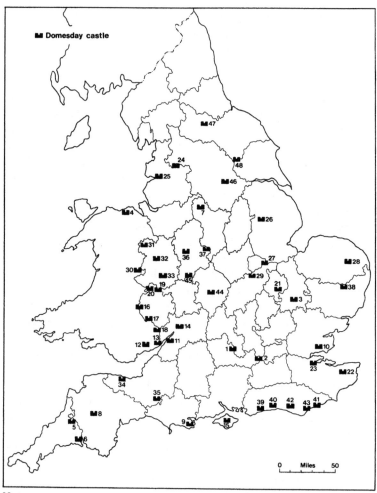

Notes

1.	Wallingford	11.	Berkeley
2.	Windsor	12.	Caerleon
3.	Cambridge	13.	Chepstow
4.	Rhuddlan	14.	Gloucester
5.	Launceston	15.	Carisbrooke
6.	Trematon	16.	Clifford
7.	Peveril	17.	Ewias Harold
8.	Okehampton	18.	Monmouth
9.	Corfe Castle	19.	Richards Castle
10.	Rayleigh	20.	Wigmore

21.	Huntingdon		35.	Montacute
22.	Canterbury		36.	Stafford
23.	Rochester		37.	Tutbury
24.	Clitheroe		38.	Eye
25.	Penwortham		39.	Arundel
26.	Lincoln		40.	Bramber
27.	Stamford		41.	Hastings
28.	Norwich		42.	Lewes
29.	Rockingham		43.	Pevensey
30.	Montgomery		44.	Warwick
31.	Oswestry		45.	Dudley
32.	Shrewsbury		46.	Tanshelf
33.	Stanton Holdgate		47.	Richmond
34.	Dunster		48.	York

Source: Based on H.C. Darby, *Domesday England* (Cambridge University Press, Cambridge, 1977), p. 316, by permission of the Cambridge University Press.

or bank and its own ditch.[5] There were a few castles erected of stone at this time, most notably the White Tower of the Tower of London, but these were very much the exception rather than the rule. Earth and timber fortifications were probably made inevitable by the need for speedy fortification and the fact that, for the most part, only an un- skilled labour force was available.[6] Castles are named, or implied, in the Domesday book in connection with 48 places, (Figure 5.1), of which 27 were boroughs; of these, only two had been built before the Conquest by Norman lords under the Confessor.[7] However, the men- tion of castles in the Domesday Book is far from systematic or com- plete and by 1086 there may well have been 70 or 80 in existence, which commanded the chief towns and main routes of southern England, guarded the approaches to and from Wales along the Welsh border, and secured communications with Normandy along the coasts of Kent and Sussex.[8] In the century following the Conquest, they greatly increased in numbers and, by about 1150, there were probably hundreds of them in existence. The wooden towers of these castles have, of course, long since disappeared and the mounds that remain have often failed to be recognised as such, sometimes being confused in older historical and archaeological sources with prehistoric and Anglo-Saxon earthworks. Nevertheless, their remains, in the form of mottes, are still to be found dotted all over the country and await mapping and identification.[9] A recent survey shows that they were often substantial earthworks ranging in height from less than 5 metres to more than 10 metres, with their diameter always at least twice their height.[10]

Moreover, so considerable was the number of castles raised in England during the first century after the Conquest that, subsequently, new castles on new sites were comparatively rare.[11] In the next two hundred years, from about 1150 to 1350, a general transition took place from the use of timber to that of stone in castle-building, mostly on the same sites. The use of stone both ensured greater permanence and also provided a better defence against improved methods of siege-craft. Frequently, the wooden tower would be replaced by a stone keep and the earthworks of the bailey would be replaced by stone walls. These are known as *keep-and-bailey* castles. Where new castles were created on virgin sites, they were often much more elaborate and included such features as gatehouses, outworks and towered walls. This steady progression to stone fortification may have been somewhat retarded during the so-called 'Anarchy' of the reign of King Stephen (1135-54), when a large number of unlicensed or 'adulterine' castles was raised. The great majority of these must have been of earthwork and timber, and they were largely destroyed after 1153 when a peace treaty secured the succession of the future Henry II. These castles of the 'Anarchy', hasty and ilicit structures derived from the urgency of war,[12] have left their remains in the form of earthwork mounds dotted about the countryside and, like those of the preceding century, await mapping and classification.

The precise number of castles at the time will never be known, but a list of castles in England and Wales published in 1959 and based upon contemporary written sources, both documents and chronicles, for the period 1154-1216, shows that some 350 were recorded as active in about 1200.[13] Allowing for the inevitable inadequacy of contemporary written evidence to record every castle then existing, we may assume there were probably about 400 castles in England and Wales at the beginning of the thirteenth century. This, undoubtedly, represents a considerable reduction of the total in existence fifty years earlier, when there were more English castles extant than at any time before or since. To the first-generation Norman castles built by the Conqueror and his successors had been added others built by second- and third-generation Normans, and during the civil war of the reign of Stephen, the large number of mainly 'adulterine' or unlicensed castles.

For the next 75 years, the Angevin kings – Henry, Richard and John – closely controlled castle-building and by confiscation greatly increased the Crown's share of castles at the expense of those of the barons. But a more important factor in reducing the number of castles was the much greater cost of building in stone, which changes in

warfare and fashion increasingly dictated. Thus, as we have seen, there
were probably some 400 castles in existence in England and Wales in
1200, and it may be estimated that this figure remained more or less
constant for the next 200 years.[14] Certainly, new or redeveloped castles
were built in the thirteenth and fourteenth centuries, but their numbers
were probably balanced by those which were abandoned or destroyed.
Castle-building in the thirteenth and fourteenth centuries reached an
apex of elaboration in the form of impressive fortifications combined
with more sumptuous residential accommodation. Apart from the great
Welsh castles built during the reign of Edward I (1272-1307), such as
those at Caernarvon, Conway, Harlech, Beaumaris and Caerphilly, there
are many splendid English examples dating from this period. They
include Tutbury in Staffordshire, mainly rebuilt after 1350; Rockingham in Northamptonshire, dating from the time of Edward I; Dunstan-
burgh and Alnwick in Northumberland with their great gatehouses built
in the fourteenth century; Bodiam in Sussex built in 1385; and not
least the great royal castle at Windsor which was substantially extended
in the second half of the fourteenth century.

By the end of the fourteenth century, the great period of castle-
building was over and the true castle was in decline. In the next fifty
years, a small number of castles was built, including the so-called
'tower houses'. These were more residential extravagances than true
castles and were generally lacking in the formidable defensive strength
of their predecessors. Thus, although they are large and substantial
enough to be described as castles, they are, in many ways, more accurately described as fortified houses. Built frequently of brick, they
include such splendid examples as Tattershall Castle in Lincolnshire,
built by Ralph Cromwell between 1430 and 1450; Caister-by-Yarmouth
in Norfolk, built between 1432 and perhaps 1446; and the Leicestershire castles of Ashby-de-la-Zouch and Kirby Muxloe built by William
Lord Hastings between 1474 and 1483, and 1480 and 1484 respectively.

An excellent example of a castle which was built and rebuilt several
times is Tutbury in Staffordshire (Figure 5.2) encompassing in minia-
ture virtually the whole history of castle-building. The original castle, a
motte-and-bailey fortress, presumably a timber and earthwork construction, was built shortly after the Conquest by the Crown on a strong
position overlooking the Dove valley. First granted by William I to
Hugh d'Avranches when he created him Earl of Chester in 1071, he
subsequently gave it to his favourite, Henry de Ferrers, who was
holding it at the time of Domesday. Ferrers, who used the castle as the

administrative centre of his extensive estates, founded the church and priory in the town below. The castle was besieged by Henry II in 1174 and although de Ferrers made his peace with the king, it seems to have been demolished at this time. Soon after, it was rebuilt in stone and the hall and chamber in the middle of the bailey belong to this period. It subsequently came into the hands of Edmund, Earl of Lancaster, who built the gateway to the north in 1313 to 1314. It was largely demolished after Edmund's rebellion against the Crown in 1322 and later came once more into Lancastrian hands, being rebuilt by John of Gaunt after about 1350. In 1399, it came with the Duchy of Lancaster into royal hands and in the fifteenth century the Crown built extensive curtain walling and the north and south towers. By the end of the Middle Ages it seems to have fallen into decline, although during the sixteenth century it was still considered a suitable prison for Mary, Queen of Scots. It was finally reduced by parliamentary forces after a three weeks' siege in 1646. Some subsequent building took place on the site, including a mock ruin called 'Julius's Tower' which crowns the motte.[15]

The growth and decline of Tutbury is mirrored in the spectacular ruins of many other English castles. The reasons for the decline of the English castle are many and complex, and reflect the changes in society which took place in the later Middle Ages. Just as the creation of Norman feudalism brought the castle into being, so the decline of the feudal system in the late fourteenth and fifteenth centuries led to its demise. The most popular reason generally advanced for the decline of the castle is the introduction of gunpowder but, according to Brown, this was not the immediate cause, as gunpowder did not render medieval fortification obsolete until the sixteenth century and, in any case, fortresses were later designed to resist and to be defended by heavy cannon.[16]

Another factor in the disappearance of the castle was the financial insolvency of the Crown. The fifteenth-century Lancastrian kings owned the majority of castles and their resources were simply inadequate to maintain them all properly. Moreover, as royal administration became more complex and bureaucratic towards the later Middle Ages, so it inevitably became more sedentary. Thus, the king settled down in and around London and the pace and extent of the royal progress through the country declined and the need for castles became less urgent.[17] Moreover, by the fifteenth century their military value was much less marked, because of changes in warfare which was being waged increasingly by larger, more professional armies that conducted

Figure 5.2: Tutbury Castle, Staffordshire

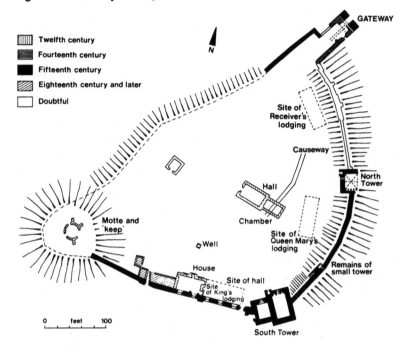

Source: Based on L.M. Cantor, 'The Medieval Castles of Staffordshire', *N. Staffs. Journal of Field Studies*, vol. 5 (1966), p. 39, by permission of the editor.

battles in the field. Finally, the gradual development of trade and industry was bringing into being a middle class which did not live in castles, and a more sophisticated aristocracy which was becoming accustomed to higher standards of comfort and residence than a castle could provide. These social and economic developments found expression in the building of fortified manor houses which had, in any case, co-existed with castles for centuries.

The fortified manor differed from the castle in that it was smaller and its fortifications were less formidable. Indeed, in the later Middle Ages at times and places when the countryside was less lawless, so the degree of fortification might be relatively slight. Fortified houses go back to the Conquest and are recorded in the Domesday Book as *domus defensabiles* (defensible houses), two examples being given, at Ailey and Eardisley in Hereford.[18] They began to develop in some numbers from the late twelfth century onwards, but were mainly built between 1270 and 1370, often in more pleasant and accessible locations than the castle.[19] Among the earliest and most beautiful examples are Stokesay in Shropshire dating back to the 1270s and Acton Burnell, also in Shropshire, built in the 1280s for Robert Burnell, Bishop of Bath and Wells.

In order to fortify his manor house, the lord was required, at least in theory, to obtain a licence to crenellate as the licensing and control of fortification was one of the most ancient and jealously guarded prerogatives of the Crown.[20] In the border counties, the Lord Marchers seem to have been empowered to grant licences to fortify houses in the areas they controlled and, in the early Middle Ages, permission to issue such licences seem also to have been vested in other territorial lords. By the middle of the thirteenth century, these licences had assumed a standard formula whereby the licensee was allowed to strengthen his manor house with a moat, and a wall of stone and lime, with crenellations and battlements. According to Parker, very few houses of any importance were built in the thirteenth and fourteenth centuries without being fortified.[21] However, the requirement to obtain a royal licence to fortify seems to have been less rigidly enforced by the Lancastrian kings between 1399 and 1470 when many fewer licences are recorded in the royal records, though this may have been due in part to a decline in the extent of building. One of the earliest licences was that granted by John, Count of Mortain in 1195 to Richard Vernon enabling him to fortify his house at Haddon in Derbyshire with walls twelve feet high without crenellations.[22] The tower of the present-day Haddon Hall subsequently acquired crenellations in the fourteenth century, when presumably a royal licence was obtained. Haddon Hall is thus a splendid example of a fortified house and one which inspired the normally laconic Nikolaus Pevsner to enthuse that, 'Haddon Hall is . . . not the forbidding fortress on an unassailable crag, but the large, rambling, safe, grey, lovable home of knights and their ladies, the unreasonable dream-castle of those who think of the Middle Ages as a time of chivalry and valour and noble feudings'.[23]

In fact, as we have seen, Haddon Hall was never a true castle in the sense of being a fortress. However, the confusion is understandable in that a number of fortified houses were entitled 'castles', though Haddon was not among them. There are, for example, three such houses in Staffordshire, all erected at different times for different owners.[24] Caverswall Castle, just to the east of Stoke-on-Trent, was a fortified house on a mound and surrounded by a moat and was built by a local lord, William de Caverswall, who was given a licence to crenellate in 1275; Eccleshall Castle belonged to the Bishops of Coventry and Lichfield and dates from about 1200 when Bishop Muschamp was given a licence to fortify by King John; and Stourton Castle, in the south-western corner of the county, was built near the end of the Middle Ages in about 1446 when John Hampton, 'esquire of the king's body' was lord of the manor.

Similar examples are found all over the country and they illustrate the variety of ownership of fortified houses. They became the principal residences not primarily of the great lords, lay and ecclesiastic, but more commonly of the newly emerging landowners, with both social pretensions and political ambitions.[25] With the agricultural prosperity of the thirteenth and early fourteenth century the wealth of this aspiring class grew and, as considerations of breeding and life-style were always of cardinal importance to them, they sought to emulate those above them by building imposing residences.[26] Judging by the numbers of licences to crenellate which appear in the royal records, principally the Patent Rolls, fortified houses reached their peak in the first quarter of the fourteenth century, a period of marked civil lawlessness. The reign of Edward II (1307–27) was marked by political unrest and rebellion and it can be no coincidence that the resultant breakdown of law and order was reflected in one of the highest concentrations of licences taken out in the Middle Ages. Indeed, the fundamental reason for the landowner to fortify was to ensure protection against the endemic violence of the medieval countryside when there was little or no help forthcoming from the authorities.

The precise number of fortified houses is not known, partly because of the difficulty in classification and partly because they have not all been recorded in documentary sources or mapped on the ground. As we have seen, the borderline between the fortified house and the true castle is a hazy one and the same is true in respect of moated homesteads, which are described later in the chapter. If one goes by the numbers of recorded licences to crenellate, then there were at least 450 fortified homes built in the Middle Ages, though there were probably many more in existence for which licences were never obtained or recorded.

In describing the place of castles and fortified houses in the medieval landscapes, two other important aspects need to be considered: their distribution and their function. As regards their distribution, they were to be found in every county in England, including not only Northumberland and those along the Welsh Marches, where they faced external enemies, but also far inland where no such threat existed: in

Staffordshire, for example, there were at least 24 with another eight possible sites, in Leicestershire 14 definite sites and another possible twelve and in Bedfordshire 22.[27] Within individual counties, their siting and distribution reflected local and regional territorial considerations, rather than any national plan. Although some castles had been erected by the Conqueror after 1066, to command the chief towns and main routes of the country and to secure communications with Normandy, these were the exception rather than the rule. The majority of castles were built by barons and other lords as residences and administrative centres. In other words, they reflected above all the baronial distribution of territory. Within a given region or county, the castle was situated on a specific site chosen for a variety of reasons, some of them conflicting If their distribution was largely determined by landholdings and the balance of power among the king and the larger landowners, nevertheless, particular sites were skilfully chosen, for tactical, strategic, economic and administrative reasons, by men with an eye for the country. Where possible, advantage was taken of natural eminences: in Leicestershire, for example, the castles at Belvoir, Castle Donington, Mountsorrel and Whitwick were built in prominent positions.[28] The castle might also be located to guard important routes as in Staffordshire, at Newcastle under Lyme, where the castle commanded the route northward to Chester, and at Tubury where the castle overlooked the Dove valley. In many cases, both purposes were served, as at Corfe in Dorset where the royal castle was built on the highest point of a hill, guarding the gap which leads through the hills of Purbeck.[29]

As we have seen, therefore, the main function of the earliest castles was essentially a military one, to hold down the conquered territory which William I distributed among his barons. Thus its prime significance was as a stronghold and garrison centre. This applied to the castles of the Crown and of the great territorial lords whose writ, at times in the Middle Ages, ran over extensive areas of the country.

The second major function of the castle was to act as an administrative centre for lands and estates belonging to the Crown or, more frequently, some other lord. Medieval society was characterised by a marked decentralisation of power from the king to his great lords and though them to lesser lords. After the Conquest, when William I granted land to his barons, they raised castles and fortified houses as administrative centres and symbols of their lordship.[30] The great Norman barons, the tenants-in-chief of the king who legally held all the land of England, allowed lesser lords in turn to hold land off them and to build castles and fortified residences. It is thus not surprising that virtually throughout the country, baronial castles outnumbered royal ones. Indeed, at different periods during the Middle Ages, individual great barons came close to rivalling the Crown in respect of the number of castles which they held, as in the case of the Dukes of Lancaster in the late fourteenth century.[31]

Partly, perhaps, castles were regarded, like monastic settlements, as instruments of colonisation and cultivation, as bases from which new land could be brought under control.[32] Partly, too, they catered for the baronial love of hunting, as centres from which the lord could ride out and indulge his passion for the chase in nearby forests, chases and parks. Indeed, some castles, like Gillingham in Dorset and Odiham in Hampshire, began as royal hunting lodges, subsequently fortified by King John, while many had parks immediately adjacent to them as at Tutbury, Nottingham and Northampton.

In conclusion it can be said that of all medieval buildings, the castle was the most characteristic of the period to which it uniquely belonged. In the words of R. Allen Brown, the castle was the

> substance of much miliary and therefore political power, the residence of the great, the cherished symbol of status and often nobility, the hub of administration and the centre in so many ways of public and private life, affecting one way or another most ranks of society through its manifold functions and the labour and service of its maintenance.[33]

Moated Homesteads

Moated sites were a very frequent feature of the English medieval landscape, numbering well over 5,100, and were to be found in every county in the country.[34] However, the definition of the term 'moated site' is very complex and the simplicity of the term is deceptive in that the only unifying feature is the moat itself. Thus, in the broadest sense, it includes a wide range of sites from castles with moats, through moated fortified houses like Stokesay Castle, to much more humble moated farmsteads. It is here taken to mean encircled manor houses and farmsteads which, in any case made up the great majority of moated sites. Indeed, it was only the presence of an encircling ditch which distinguished them from the normal manor house. Their owners were essentially the smaller feudal landowners and their moated homesteads were working units in the land-use pattern of the time.[35]

In straightforward archaeological terms, a moated homestead may be defined as an area of ground occupied by a dwelling, often with associated structures, bounded or partly bounded by a ditch which, in most cases, was intended to be filled with water.[36] From an archaeological point of view, one of the problems in identifying moated homestead sites is that they can easily be confused with other features of the medieval landscape, especially fish-ponds, many of which are known as moats and are sometimes so described on maps, perhaps because they were commonly associated with moated sites. Moreover, many medieval moated sites have been subsequently modified for other purposes such as ornamental ponds or for stock watering or have been turned into formal gardens.[37] In recent years, the filling in of all or part of the moat has become increasingly common. For all these reasons, moated sites are by no means easy to identify.

However, they have in common certain basic physical characteristics. The typical moated site consists of an 'island' of roughly rectangular form, surrounded by a ditch. Inevitably, there are many exceptions to the rule and they may, on occasion, be oval, circular, triangular, pentagonal or irregular in shape, as well as having a 'double' form with a smaller moat with a larger enclosure.[38] The enclosing ditches vary greatly in size but are usually between 3 and 6 metres across. The 'island' may be about 4,000 square metres in size with sides approximately 250 and 160 metres respectively,[39] usually located on flat land or on land level with the surrounding countryside. The size of the dwelling built on the site varied considerably, from a quite small and simple house to a much larger one with yard, barns, stables and gardens. Occasionally, complex sites are found with two or more moated enclosures, one occupied by the house, the other by various outbuildings and gardens. Access to the moated homesteads was provided by a causeway, drawbridge or permanent bridge.

Although they were clearly a common and characteristic feature of the medieval landscape, the precise functions and social and economic significance of moated homesteads can only be outlined, as a full and definitive account will have to wait upon more detailed research. However, much of what is presently known about them derives from the painstaking work of the Moated Sites Research Group under their secretary, Mr F.A. Aberg. First and foremost, they must be regarded as dwellings and were probably to be found quite early in the Middle Ages. Certainly, they were in existence by 1150 and appear to have reached their apogee in the period 1200-1325. Although they were not heavily defended, they would nevertheless have provided a modicum of protection in a lawless age and, perhaps what was more important, also supplied an element of psychological comfort.[40] They also provided protection in two other ways. First, the moats made available a copious supply of readily accessible water in case of fire, an ever-present hazard in the Middle Ages when timber and thatch were the usual building materials. Second, they were a protection against the depredations of wild animals, such as wolves and foxes, which might carry off young livestock, and deer, which were shielded by the forest law and voracious eaters of stored crops, hay and garden produce.[41] An obvious motive for building moated homesteads in some parts of the country was to facilitate drainage, a considerable advantage, for example, in the winter months in the clay lowlands of the Midlands. The moat itself was put to a variety of uses: it could also be used to water livestock, especially in winter; it could be stocked with fish, a valuable form of protein; and it could be used as a swannery.

Finally, a moated site must also, at different times and in different places, have constituted a status symbol. It seems likely that it would originally have been the dwelling of the aristocratic classes but, as the country became more prosperous in the thirteenth and early fourteenth centuries, so it spread downwards through the social scale into the upper levels of the peasantry, lesser knights and freemen. Indeed, as a result of a detailed study of Cambridgeshire moated sites, Taylor suggests that the owners of moated homesteads may not have built them primarily for material reasons, but rather to show off their property by imitating the higher ranks of contemporary society.[42] In doing so, they were heirs to a tradition of moat construction that antedated the Conquest. Certainly, the moat itself can be seen as a symbol of feudal society, emphasising as it did the separateness and exclusiveness of its owner.[43]

As we have seen, there were well over 5,100 moated sites in England during the Middle Ages, of which the great majority were moated homesteads. In terms of their chronological development, Le Patourel and Roberts identify four main phases.[44] The first was during the century following the Conquest, when the first medieval moated sites were constructed and the form began to evolve. The second phase occurred between 1150 and 1200 when they were becoming more widespread and were to be found, albeit in fairly small numbers, throughout the country and against a wide variety of topographical backgrounds. During this period, they may well have been more defensive in character than was the case later, especially during the anarchic reign of King Stephen and the troubled times of Henry II (1135-89).

It was during the third phase, between 1200 and 1325, that moated homestead construction reached its apogee and probably about 70 per cent of the known sites came into being during this period. As we have seen, this was a time of agricultural expansion and population growth which also witnessed the greatest development of the medieval park. Finally, as with the park, the fourth phase was one of decline, following the plagues of the mid-fourteenth century and the economic recession of the fourteenth and fifteenth centuries. By the end of the Middle Ages, the creation of new wealth from sheep farming and commercial activities enabled local gentry to build new houses on a larger, more grandiose scale and to forsake what Roberts calls 'the damp, insanitary sites of former ages'.[45] However, not all moated sites were unhealthy and damp and, as Lord Hastings's splendid, if uncompleted, fortified house at Kirby Muxloe in Leicestershire, testifies, the construction of at least one late fifteenth-century grandiose house was undertaken within a moat.

An analysis of the distribution of moated sites, both nationally and on a county-by-county basis, reveals some interesting patterns and correlations. The Moated Site Research Group (MSRG) has identified 5,307 known sites in England, Wales and Scotland, of which 5,140 are in England, 136 in Wales and 31 in Scotland. They are found in every county in England, with the greatest concentrations in Essex (548), Suffolk (507), Yorkshire (320), Lincolnshire (297), Cambridgeshire (270) and Worcestershire (232). On the other hand, six counties had fewer than 20 known sites — Cornwall, Cumberland, Devon, Northumberland, Rutland and Westmorland. Once again, the similarity between the relative distribution of moated sites and medieval parks is striking. However, any conclusions must be tentative as many more moated sites await discovery and, in any case, the present list of known sites reflects in part the relative intensity of the work of members of the MSRG and other field-workers. The location of the known sites has been recorded on a distribution map of England and Wales,[46] and an examination of this map reveals an obvious and important link with lowland areas underlain by clay, in northern, central and eastern England (Figure 5.3);[47] in other words, moated sites appear most frequently in those parts of the country where free-draining soils, such as those underlain by chalk, limestone, sandstone and alluvial gravels, are conspicuously absent. Over the country as a whole, three areas emerge as having the greatest concentration of sites: the tract of country between Chelmsford and Harlow in Essex and Bishop's Stortford in Hertfordshire; an area extending from Suffolk into north Essex, with a

Figure 5.3: Moated sites in England and Wales

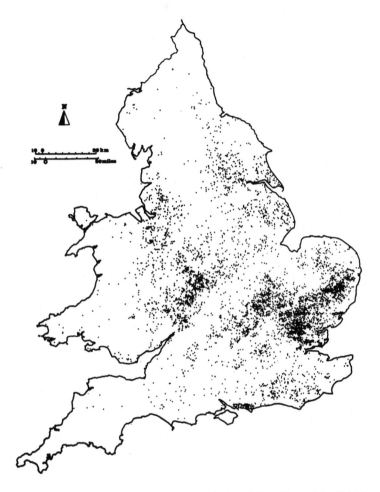

Source: Based on F.A. Aberg (ed.), *Medieval Moated Sites* (Council for British Archaeology Research Report, No. 17, London, 1978), p. 2, by permission of the editor.

particular concentration north of Framlingham; and the Birmingham area extending south into Warwickshire and Worcestershire. On the other hand, the more geographically extreme parts of the country, to the north and to the south-west, contain virtually no moated sites as do such areas of inhospitable high ground as the Pennines and Exmoor.

If we look more closely at the distribution of moats within individual counties, we find that here, too, geology plays an important part though other, more local, factors are also at work. For example, in many counties moated sites are thickly clustered in former areas of royal forest which were cleared and brought under cultivation during the Middle Ages, at a later period than the main Anglo-Saxon settlement. This is well illustrated in the case of Worcestershire, where they are particularly concentrated in the medieval extent of the Forest of Feckenham and Malvern Chase, where much assarting took place during the twelfth and thirteenth centuries.[48] On the other hand, they are markedly absent in the northern part of the county in the Forests of Wyre and Kinver which remained as royal forests throughout the Middle Ages and were thus protected from assarting. Within the county, there is also a relationship between the distribution of moated sites and soil types. Thus, the maximum density occurs in a broad east–west belt across the middle of the county which can to some extent be correlated with the Keuper Marl and Lower Lias Clay plateaux which generally provide heavy, ill-drained soils. By contrast, in the south-eastern part of the county, the fertile soils of the Vale of Evesham were very early colonised with nucleated villages and open-field cultivation which largely precluded moated sites.

A similar pattern can be observed in the adjacent county of Warwickshire, with its twofold division into Arden and 'Fielden' Warwickshire, respectively north and south of the river Avon. The former was poor and infertile, well-wooded and settled relatively late, while the latter was prosperous with fertile soils and little woodland, and settled in the Anglo-Saxon period. It is significant therefore that north and east of the Avon the moated sites are plentiful, while to the south they are very few and far between. As Roberts has pointed out, the distribution of moats and primary Anglo-Saxon settlement names in Warwickshire is, therefore, almost completely mutually exclusive.[49] A similar pattern of distribution also occurs to the west and east of the lower Severn between Gloucester and Tewkesbury.[50]

In Essex, however, the county with the largest recorded number of moated sites, the relationship is more complex.[51] The north-western part of the county was, at the time of Domesday, most densely settled with the highest number of plough teams, yet moated sites became relatively frequent. However, small particles of woodland were common here, and their clearance in the Middle Ages for the purposes of cultivation may well have been accompanied by the construction of moated homesteads. On the other hand, the south-western part of the county was royal forest throughout the medieval period and, as one would expect, contains a small number of moated sites. Finally, the lower parts of the county along the river valleys and the coastal marshes are characterised by a marked absence of moated sites, which may be due to the fact that this land was mainly pasture during the Middle Ages, the homesteads being situated in areas of more intensive land use.

Last, there may well be a correlation between the distribution of moated sites and feudal land tenure. In Lincolnshire, for example, where moated homesteads were very common, much of the land was held by socage, that is by freemen in exchange for fixed services to the lord, so that large manors centred on a single village were rare. As a result, the tenurial framework of land-holding was such that moated settlements would easily develop, and Emery instances examples of an exact correlation between the five or six Domesday sokemen with free estates in 1086 and the number of moated settlements surviving today.[52] He also makes the point that the rapid growth in population in the mid-thirteenth century led to a demand for land by a large class of freeholders which had recently emerged. The larger landowners, impoverished by the Barons' Wars, met their demands by subdividing their manors and granting lands by subinfeudation to freeholders. Many of this new class of subtenants would thereupon have improved an existing hall or built a new house and safeguarded their property with a moat.

Clearly, the pattern of distribution of moated sites, whether on a national or county basis, is the product of a complex interweaving of a wide range of social, economic, historical and geographical factors. Their contribution to the medieval landscape was a significant one and a more detailed analysis of that contribution will have to await further study on the ground. This makes it all the more important to record the characteristics of present-day moated sites which, like all other historical monuments, are in increasing danger of being obliterated by modern farming methods and urban and other developments.

Monastic Settlements

The part played by the monastic orders — both for men and women — in the religious, social and economic life of medieval England was, of course, enormous and has been well documented.[53] The monastic orders, notably the Cistercians, helped to transform the landscape by introducing agricultural innovations such as sheep farming, draining the marshland, and clearing the wasteland. Through their contacts with the Continent they helped greatly to increase trade and to foster economic development. These aspects of their influence on the English medieval landscape are dealt with elsewhere in this book; here we are concerned with the settlements they founded, their character and distribution.

During the Middle Ages, almost every large building in the English landscape that was not a castle or manor house was almost certain to be a religious house of some kind. In all, the religious orders founded over one thousand monasteries or houses. They were of various sizes, scattered throughout England and situated both in towns and in the countryside. The large monasteries were usually known as abbeys, and other houses were classed as priories. The latter were often as large and wealthy as the greater abbeys, especially cathedral priories in towns like Durham and Winchester.[54] The term 'priory' was also applied to daughter houses established by monks from an abbey and to monasteries that were dependants of larger houses in England or the Continent, the last-named being known as 'alien priories'.

Religious houses were already well established in England at the time of the Conquest, mainly within the built-up areas of towns large and small, such as Canterbury, Winchester, London, Gloucester and Bury St Edmunds. These were all Benedictine foundations and, by 1066, England contained about 50 of them.[55] Though they were scattered throughout the country, there were concentrations of monasteries in three main areas: in the Fenlands, including the great abbeys of Peterborough, Ely, Ramsey, Thorney and Crowland; in the Seven Valley, around Gloucester, Tewkesbury and Pershore; and in the southern counties of Wiltshire, Dorset and Hampshire, at Glastonbury, Malmesbury, Bath, Cerne, Sherborne and Winchester (Figure 5.4). They were nearly all richly endowed and possessed about one-sixth of the land of England.[56]

The Conquest was not followed by the creation of many new religious houses, as William I, though personally pious and austere, kept a tight hold on the church. However, a number of new monasteries were set up in the years after 1066: William himself founded Battle Abbey on the site of his victory; some of his barons created monasteries in provincial towns such as Shrewsbury, Chester, Colchester and Lewes; and some scores of minor houses were founded in various parts of the country. Thus, by the end of the eleventh century, the number of religious houses in England had risen to about 130, of which 45 were independent abbeys and another 45 were alien priories.[57] All these

foundations were Benedictine, except for five Cluniac houses, which were subject to the reformed Benedictine abbey of Cluny in Burgundy, at Bermondsey, Wenlock, Castle Acre, Pontefract and Lewes. The Cluniac order, which laid great stress on worship and required an elaborate ritual, spread slowly in England and by 1160 owned about 36 houses, of which only about a third were full-scale priories, the rest consisting of small cells, where two or three monks led lonely lives.[58]

In the twelfth century, the structure of English monachism became more complex and a wave of enthusiasm began to fill the country with religious houses. The dominant position of the Benedictines was overtaken by the growth and popularity of other Orders that put down fertile roots into English soil, particularly the Cistercians, or White Monks. Founded by an Englishman, Stephen Harding, and based on the Abbey of Citeaux, in Burgundy, they aimed to bring back the simplicity and hard labour which the Benedictines had abandoned. Accordingly, they turned their backs on the towns and adopted a rule that the sites of their abbeys should be far from the habitation of man. Consequently, they colonised remote and unpopulated areas in the Midlands, north and west of the country, building abbeys in such places as Bordesley in Worcestershire and Stoneleigh in Warwickshire, on the edges of the Forests of Feckenham and Arden and, above all, at Fountains, Rievaulx and Kirkstall in Yorkshire. The first Cistercian house was founded in 1128, at Waverley in Surrey; within twenty-five years of their arrival 50 Cistercian monasteries had come into being, and by the end of the twelfth century over 230 abbeys and priories had been established.[59] Their remarkable economic success has been attributed to a number of factors: although their houses were located in remote places, the Cistercians, through their dedicated labour, made the land blossom into wealth; they introduced profitable agricultural practices, such as sheep farming; they welcomed illiterate and humble followers who became the field-workers upon whom their material prosperity depended; and they succeeded, during their early years at least, in avoiding the payment of substantial taxes to the Crown.[60]

One of the most picturesque and celebrated of the Cistercian houses was the Abbey of Fountains, near Ripon in the West Riding of Yorkshire which Pevsner describes as 'the pattern in England of the arrangement of the monastic quarters of a large Cistercian house'.[61] The first prerequisites for such a monastery were a fairly level site to allow for the construction of the basic buildings and a water supply for drinking, washing, drainage and fish-ponds. In the case of Fountains Abbey, it was located on the level floor of the narrow and shallow valley of the river Skell, four miles south-west of Ripon (Figure 5.5 and Plate 5). Building began here in about 1135 and was completed by about 1250. Nothing further of any great note was added until the North Tower was erected shortly before the abbey was dissolved in 1539. In architectural terms, therefore, Fountains Abbey possessed a homogeneity of style whose ruined splendour leads Pevsner to enthuse 'there is no other place in the country in which the mind can so readily evoke the picture of thirteenth century monastic life'.[62]

Entry to the abbey was gained by the great gate which lay to the west. Although only fragmentary ruins remain today, the gatehouse would have been set in a massive stone ring-wall; it would have been vaulted and high enough to permit carts to enter. Inside the precinct, east of the bridge, lay the great houses which accommodated travellers and fed the poor. From here, the visitor approaching the conventual buildings would be faced by what was, in effect, a castle-like inner bailey, which at Fountains was made up of the west end of the church and the western range of the cloister. The major and biggest building was the abbey church which was about 360 feet long with a tower rising to about 170 feet. To the south lay the west range of the cloister which comprised a cellar below and a dorter, or monk's dormitory, above. The cloister itself provided study accommodation for the monks and lay brothers. Other conventual buildings included a noble vaulted chapter-house where the abbot met to hear confessions and transact monastic business, a refectory, an infirmary and the abbot's lodging. These and other buildings made up a great architectural complex whose remains today comprise one of the most famous and most beautiful monastic ruins in the whole of England.

In addition to the Cistercians, the twelfth century also witnessed the emergence of other popular Orders, especially in the shape of the Canons Regular. They were members of three Orders, Augustinian, Premonstratensian and Gilbertine, who formed small communities of priests who had taken monastic vows and led the normal lives of monks, or regulars, except that they were ordained. Between them they established some 230 houses, dotted all over the country (see Figure 5.4).[63]

The Benedictines and Cistercians both adopted the principle of enclosure within monastic precincts, in contradistinction to two military Orders, the Knights Templars and Knights Hospitallers, who were founded early in the twelfth century after the first Crusade. They established a number of houses in England whose main functions were to train new members and to provide retirement for those whose

Figure 5.5: Fountains Abbey

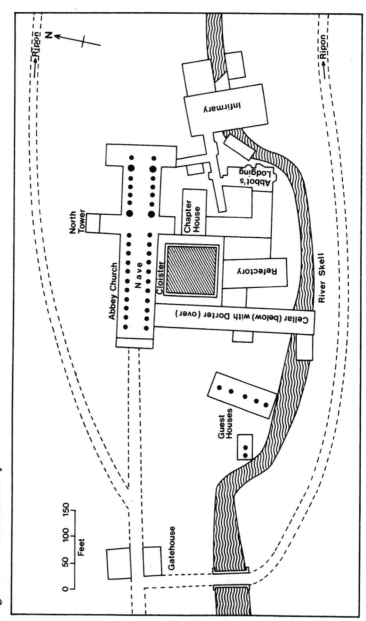

Plate 5: Fountains Abbey, near Ripon, North Yorkshire

fighting days were over. Once the Crusades were at an end, by the end of the twelfth century, the Templars lost their *raison d'être* and the Order was finally abolished in 1312. By that time, they possessed 20 or more houses in England, most of which were made over to the Hospitallers,[64] who had eventually had about 50 commanderies in Great Britain,[65] which survived until they were dissolved in 1540. The Knights Hospitallers ran their estates much like other medieval landowners and a clear picture of their economic management in 1338 emerges from contemporary documents.[66] In each manor, the estate was managed by individual brethren, or confratres, and was serviced by a wide range of chaplains, chamberlains, squires, foresters and servants in the house and stables. The principal charge was on the maintenance of the estate and the provision of hospitality for the poor. The largest item of expense on food was on cereals, very large amounts being used for baking and brewing — three types of bread and two classes of beer were commonly produced — and for comsumption in the stables. In addition, substantial quantities of flesh and fish were required. The chief source of income was derived from the keeping of sheep which was extremely profitable due to the high value of wool. As a result large flocks were kept; for example, the Hospitallers' manor of Hampton in Middlesex owned 2,000 sheep. In addition, pasture was loaned out to other owners of sheep, at the normal rate of one penny a head of sheep; thus, Beverley and Dalton, both in the East Riding of Yorkshire, pastured 480 and 240 sheep respectively, and Ossington in Nottingham and Chippenham in Wiltshire each pastured 600. The dovecote seems also to have yielded a high return in some manors and was a steady source of profit.

By the fourteenth century, although some monasteries had become wealthy, partly as a result of their own efforts and partly through endowments, many others, because of declining benefactions and taxation, were increasingly subject to financial distress. Moreover, there was a decline in the number of men and women coming forward to be monks as the opportunities for literate men and women outside the monasteries increased in a more sophisticated society. This decline became catastrophic with the arrival of the Black Death in the middle of the fourteenth century. This was the greatest calamity to befall the monasteries prior to the Dissolution; it halved the number of monks and regular canons and visited the principal monastic centres of the Benedictines and the Cistercians with extreme severity. Although there was a revival in monasticism after 1360, the numbers of monks in 1400 were still only about two-thirds of what they had been in the early thirteenth century.[67] Further outbreaks of plague in 1361 and 1368 made the position of the Orders increasingly difficult and they were never again as influential and widespread as they had previously been. As farm labourers were no longer available, so Cistercian abbots, for example, like other great landowners, had to divide up their large farms into smaller ones that could be farmed out on short leases. Moreover, the growth of nationalism and anti-Papal feeling in the country had manifested itself in the confiscation by the Crown of a number of alien priories, a process which was completed in 1414 when all the remaining ones were taken over. By the time of the Dissolution, a large number of small monasteries and nunneries had fallen into decay and many others had greatly reduced numbers of inmates; as a result, there remained about 650 religious houses of the various Orders in England and Wales.[68]

The contribution of the monastic Orders, in terms of settlements, to the English medieval landscape was thus a substantial and varied one. Their houses were to be found, in very substantial numbers, throughout the country, both in towns and cities and in areas remote from habitation. As can be seen from Figure 5.4, no county was without them, though they were particularly numerous in the Midlands and East Anglia and, thanks to royal dispensation, virtually every forest in England contained one or more monastery.[69] The direct influence of the Church upon the English landscape, not least where buildings were concerned, was therefore a very substantial one. It was reduced after the middle of the fourteenth century, largely as a result of the plagues, but was still very considerable until the Dissolution, which suppressed the monastic Orders and reduced their many splendid buildings to relict features so familiar in our landscape today.[70]

Notes

1. Christopher Hohler, 'Court Life in Peace and War' in Joan Evans (ed.), *The Flowering of the Middle Ages* (Thames and Hudson, London, 1966), p. 157.
2. R. Allen Brown, *English Castles*, 3rd rev. edn. (Batsford, London, 1976), p. 14.
3. Hohler, 'Court Life in Peace and War', pp. 162–3.
4. See, for example, R. Allen Brown *et al.*, *The History of the King's Works*, vol. I, *The Middle Ages* (HMSO, London, 1963); D.F. Renn, *Norman Castles in Britain* (John Baker, London, 1970); and S. Toy, *Castles of Great Britain* (Heinemann, London, 1953).
5. H.C. Darby, *Domesday England* (Cambridge University Press, Cambridge, 1977), p. 313.
6. Brown, *English Castles*, pp. 53–4.

7. Darby, *Domesday England*, p. 314.

8. Ibid.

9. See, for example, L.M. Cantor, 'The Medieval Castles of Leicestershire', *Trans. Leics. Arch. and Hist. Soc.*, vol. LIII (1977-8), pp. 30-41.

10. D.J.C. King, 'The Field Archaeology of Mottes in England and Wales', *Chateau-Gaillard*, vol. V (1970), pp. 101 ff.

11. Brown, *English Castles*, p. 81.

12. Ibid., p. 82.

13. R. Allen Brown, 'A List of Castles, 1154-1216', *English Historical Review*, vol. LXXIV (1959), pp. 49-51.

14. Brown, *English Castles*, p. 216.

15. L.M. Cantor, 'The Medieval Castles of Staffordshire', *N. Staffs. Journal of Field Studies*, vol. 5 (1966), pp. 38-46.

16. Brown, *English Castles*, p. 221.

17. Ibid., p. 222.

18. Darby, *Domesday England*, p. 315.

19. D. Williams, 'Fortified Manor Houses', *Trans. Leics. Arch. and Hist. Soc.*, vol. L (1974-5), p. 2.

20. Ibid., p. 1.

21. J.H. Parker, *Domestic Architecture of the Middle Ages*, vol. II, pt II (OUP, Oxford, 1851).

22. N. Pevsner, *The Buildings of England: Derbyshire*, 2nd edn. (Penguin, Harmondsworth, 1978), p. 224.

23. Ibid., p. 221.

24. Cantor, 'The Medieval Castles of Leicestershire', pp. 43, 45.

25. Williams, 'Fortified Manor Houses', p. 4.

26. Ibid.

27. For Staffordshire, see L.M. Cantor, 'The Medieval Castles of Staffordshire'; and D.M. Palliser, 'Staffordshire Castles: A Provisional List', *Staffordshire Archaeology*, no. 2 (1972), pp. 5-8; for Leicestershire see L.M. Cantor, 'The Medieval Castles of Leicestershire'; and for Bedfordshire, see D.J.C. King and L. Alcock, 'Ringworks of England and Wales', *Chateau-Gaillard*, vol. iii (1966), which gives overall and county totals of castle sites in England and Wales.

28. L.M. Cantor, 'The Medieval Castles of Leicestershire', p. 33.

29. See R.R. Sellman, *Illustrations of Dorset History* (Methuen, London, 1960), pp. 24-5, which includes a distribution map showing the position of 13 Dorset castles, five royal and eight baronial.

30. Brown, *English Castles*, p. 218.

31. S. Armitage-Smith, *John of Gaunt* (Constable, London, 1904), p. 218.

32. Brown, *English Castles*, pp. 218-19.

33. Ibid., p. 221.

34. F.A. Aberg (ed.), *Medieval Moated Sites* (Research Report No. 17, Council for British Archaeology, London, 1978), p. 3.

35. F.V. Emery, 'Moated Settlements in England', *Geography*, vol. XLVII (1962), p. 378.

36. C.C. Taylor, 'Moated Sites: their Definition, Form and Classification' in Aberg (ed.), *Medieval Moated Sites*, p. 5.

37. Ibid., p. 8.

38. Emery, 'Moated Settlements in England', p. 382.

39. H.E. Jean Le Patourel and B.K. Roberts, 'The Significance of Moated Sites' in Aberg (ed.), *Medieval Moated Sites*, p. 49.

40. Ibid., p. 7.

41. Emery, Moated Settlements in England', pp. 384-5.

42. C.C. Taylor, 'Medieval Moats in Cambridgeshire' in P.J. Fowler (ed.), *Archaeology and the Landscape* (John Baker, London, 1972), p. 246.

43. Jean Le Patourel and Roberts, 'The Significance of Moated Sites', p. 48.

44. Ibid., p. 51.

45. B.K. Roberts, 'Moated Sites', *The Amateur Historian*, vol. 5 (1961-3), p. 37.

46. Aberg (ed.), *Medieval Moated Sites*, p. 2.

47. Jean Le Patourel and Roberts, 'The Significance of Moated Sites', p. 49.

48. C.J. Bond, 'Moated Sites in Worcestershire' in Aberg (ed.), *Medieval Moated Sites*, p. 71.

49. Roberts, 'Moated Sites', p. 35; see also p. 40 for a distribution map of Warwickshire moated sites.

50. Emery, 'Moated Settlements in England', p. 386.

51. J. Hedges, 'Essex Moats' in Aberg (ed.), *Medieval Moated Sites*, p. 64.

52. Emery, 'Moated Settlements in England', p. 386.

53. The great standard work on the history of monasticism in England and Wales is D.M. Knowles, *The Religious Orders in England* (Cambridge University Press, Cambridge, 1955). Other excellent shorter works are D.M. Knowles and R.N. Hadcock, *The Medieval Religious Houses of England and Wales* (Longman, Green and Co., London, 1953), which contains a comprehensive catalogue of religious foundations in England and Wales, together with distribution maps; G.H. Cook, *English Monasteries in the Middle Ages* (Phoenix House, London, 1961); and L. Butler and C. Given-Wilson, *Medieval Monasteries of Great Britain* (Michael Joseph, London, 1979),.

54. Cook, *English Monasteries in the Middle Ages*, p. 15.

55. Butler and Given-Wilson, *Medieval Monasteries of Great Britain*, p. 26.

56. Ibid.

57. Cook, *English Monasteries in the Middle Ages*, p. 45.

58. D.M. Stenton, *English Society in the Early Middle Ages, 1066-1307* (Penguin, Harmondsworth, 1959), p. 233.

59. Cook, *English Monasteries in the Middle Ages*, p. 45.

60. Stenton, *English Society in the Early Middle Ages, 1066-1307*, p. 235.

61. N. Pevsner, *Yorkshire, The West Riding* (The Buildings of England), (Penguin, Harmondsworth, 1974), p. 210.

62. Ibid., pp. 203-4.

63. Cook, *English Monasteries in the Middle Ages*, p. 117.

64. Ibid., p. 200.

65. Butler and Given-Wilson, *Medieval Monasteries of Great Britain*, p. 61.

66. 'The Knight Hospitallers in England, AD 1338', *Camden Soc.* (1855-6), pp. xviii-xxxii.

67. A.R. Myers, *England in the Late Middle Ages, 1307-1536*, (Penguin, London, 1963), p. 65.

68. Cook, *English Monasteries in the Middle Ages*, p. 15.

69. J.C. Cox, 'Forestry' in W. Page (ed.), *Victoria County History, Nottinghamshire*, vol. 1 (Constable, London, 1906), p. 373.

70. See, for example, D.M. Knowles and J.K. St Joseph, *Monastic Sites from the Air* (Cambridge University Press, Cambridge, 1962); and Butler and Given-Wilson, *Medieval Monasteries of Great Britain*.

6 VILLAGES AND TOWNS

Peter Bigmore

Villages and towns of medieval England have long been regarded as central components of the total landscape fabric, their developments inextricably intertwined and their fate controlled by the same external forces of the natural environment and of times of prosperity and recession. While the town stood largely divorced from the production of food, it relied, ultimately, upon the resources of the countryside and the fortunes of town and country would rise and fall together. However, although it would be foolish to ignore the sustained growth of economic activity in the early Middle Ages and its marked decline in the fourteenth century as a major force for change, in more recent years researchers have recognised the important role played by individuals, particularly lords of the manor, in shaping the form and fortune of the individual town and village.[1]

It is the form of towns and villages that has often attracted the attention of those interested in the medieval landscape.[2] The nucleated village clustered around the parish church and surrounded by its great open fields is often our image of the medieval rural scene. Such villages certainly did exist, but a more complex picture has now emerged involving a wide variety of forms and patterns and the stability of settlements of supposed Dark-Age origins is seriously questioned. The classic village–church relationship works well in some of the Midland counties, but breaks down in the edge of the upland zone, the Welsh Marches and the south-west, and can produce considerable problems if applied to East Anglia and the south-east. In those regions the village is but one part of a complicated rural pattern, with hamlets and isolated farms containing the majority of the population and the church standing alone in the fields. The village itself might lie scattered around a great common with no clear focus recognisable. In such cases it is often difficult to distinguish between village and hamlet; any distinction based simply upon size or form is of little value, and other attempts at an association between village and church have already been shown as unsuitable. There is perhaps some justification in the recognition of a primary centre where settlement and parish name coincide, but this again need not have taken the form of a nucleated village.

In the case of the medieval town, definition remains a matter of debate. However, an acceptable one is that of Professor Beresford who has included into his town category all those that possessed at least one of the following features: a borough charter; a mention as a 'burgus' in the Assize Rolls; the reference to burgesses; and whether it sent members to any medieval Parliament or was separately represented by a jury before the judge of assize.[3] Such tests are invariably legalistic rather than functional as they reflect the emphasis of early medieval documents and any simple definition based upon whether the settlement possessed a market would give a false impression of the nature of the medieval town. This uncertainty does create problems in the compilation of any definitive list of medieval boroughs,[4] particularly where the separation of small market town and large village is almost impossible. Throughout the medieval period London stood preeminent: although its Domesday population barely reached 10,000, only Norwich and York could achieve more than 5,000 and the great majority of Domesday boroughs had fewer than 1,000 inhabitants. Many must have appeared almost as rural as the villages they served.[5] Their distribution (Figure 6.1) emphasises the greater prosperity of the countryside south and east of the Humber–Severn line. By 1300, London had increased its lead, with 50,000 inhabitants, but Bristol had emerged as the second-order town of the hierarchy (17,000), followed by Norwich and York (7,000–8,000 each). The top end of the hierarchy was to remain largely undisturbed through the rest of the medieval period, although Newcastle (a post-Domesday foundation) and Exeter were eventually to rank with Norwich and York. Yet in 1500 there were still some 500–600 English towns in the small market-centre category, with populations between 500 and 1500.[6]

However, the great majority of English towns had a population of 1,000 or fewer and many retained a strong rural atmosphere with burgesses being involved part-time in agriculture and great open fields lying just outside the town's defences; those of Stamford, for example, were not enclosed until the second half of the nineteenth century.[7] Even within the town the medieval fabric would have differed little from that of the village, for timber-framed houses in vernacular tradition were almost ubiquitous. Yet these are but minor aspects when we come to consider the major morphological and functional contrasts that came to divide village and town in medieval England.

Figure 6.1: Medieval towns in England

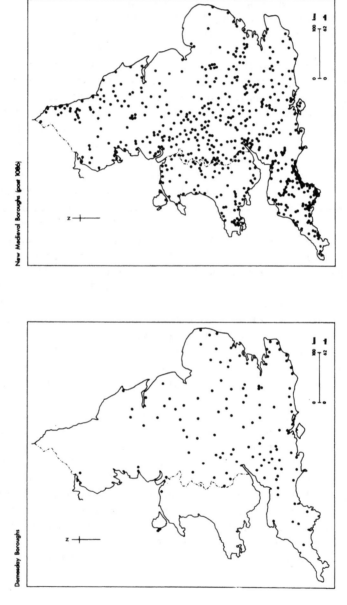

Domesday Boroughs

New Medieval Boroughs (post 1086)

Source: Based on R.A. Dodgshon and R.A. Butlin (eds.), *An Historical Geography of England and Wales* (Academic Press, London, 1978), p. 107.

Villages

The Norman Conquest and the compilation of the Domesday Book mark the traditional starting-point of this volume but their relevance to medieval settlement studies has perhaps been overemphasised. There is little evidence that the Conquest made a significant impact upon either village form or pattern, although it had resulted in the largest change-over in land-ownership, from Saxon thegn to Norman lord, witnessed in historic time. In the northern border counties, however, the Conqueror's scorched-earth policy of 1069–70 created waste over large areas and communities needed to be restructured.[8] The uncertainty of the border in Cumbria was not resolved until 1092, and then only temporarily, when William II was able to wrest control from northern lordships.[9] The Anglo-Saxon Chronicle describes how the Norman king was able to stamp his authority upon the area:

> King William marched north to Carlisle with a large army, and re-established the fortress, and built the castle, and drove out Dolfin who had previously ruled the land there, and garrisoned the castle with his men, and afterwards returned to the south, and sent thither very many English peasants with wives and stock to dwell there and to till the ground.

It seems likely that those villages which lie today between the central massif of the Lake District mountains and the Solway Firth, with their Scandinavian and English place-names, were those settled by William at the end of the twelfth century. Elsewhere, the creation of new villages was not to be repeated, and in the south communities were sometimes removed as the Norman kings made huge extensions to their royal hunting preserves. The establishment of the New Forest meant the destruction of a number of communities, although the poor sandy soils that cover much of the area would have been unlikely to support a large population and we might not be wise in looking for large nucleated villages.[10]

Professor Darby's seminal work on the Domesday geography of England[11] has awakened our senses to the great wealth of detail that is to be found within the Domesday Book, and warned us of some of the pitfalls to be encountered in the study of rural settlement. Too often, researchers have regarded the evidence of the Domesday Book as almost unequivocal in its description of the English village and as the earliest documented source for most place-names it has been used in

error or misinterpreted. Recent estimates suggest that only 5 per cent
of the Domesday population were town-dwellers and as the survey was
aimed at covering the great landed estates rather than the wealth of
the towns, it is hardly surprising that most research has been concerned
with the countryside.

One of the greatest errors has been in the assumption that because a
present-day village can be found named in the Domesday Book, and
that the name is of Scandinavian or English origin, then the contemporary village lies over the site of a Dark-Age predecessor. This may
well be true for some villages, but the excavation of such early Dark-
Age sites as West Stow (Suffolk),[12] Chalton (Hampshire),[13] Bishopstone (Sussex),[14] Mucking (Essex)[15] and Catholme (Derbyshire),[16]
where substantial Anglo-Saxon settlements lie some distance from the
present village, suggests that an alternative hypothesis needs to be
sought. The technique of field-walking, a systematic coverage of
ploughed fields that will reveal scatters of pottery and other evidence of
settlement debris, has yielded rich rewards in the central parishes of
Northamptonshire that appear to confirm the evidence of the larger
excavations.[17] Some parishes revealed several hamlets of early Saxon
date, none of which lay close to the existing village. Who is to say
which one, if any, was the predecessor or even bore the same place-
name? It is from such evidence as this that a growing body of opinion
has emerged that recognises the first two centuries of the Dark Ages
not as a period of stability in which the first of our present-day villages
were established but one of high mobility, of short-distance migration
within a framework of more ancient estate boundaries. If number of
houses is any criteria then this 'Mid-Saxon Shuffle' involved small
hamlets rather than large villages.

Although the Domesday Book provides a copious amount of detail
on rural settlement, it is not given in a useful form as far as understanding what the medieval village looked like. The entries are organised
on a basis of land-ownership and manors rather than settlements.
Where a place-name appears more than once in the folios (and this is
the case for the majority of Domesday place-names) then it is almost
impossible to decide whether the details relate to one or more settlements (Figure 6.2). Later medieval documents are also unhelpful in
this context, for taxation lists and manorial documents will do little
more than list place-names of the major vills or manorial centres and
to ignore the isolated farms, hamlets and, perhaps, nucleated villages
that formed part of a larger estate. Professor G.R. Jones has argued
cogently for the recognition of an administrative system of multiple

Figure 6.2: Great, Little and Steeple Gidding, Huntingdonshire: the fission of medieval parishes

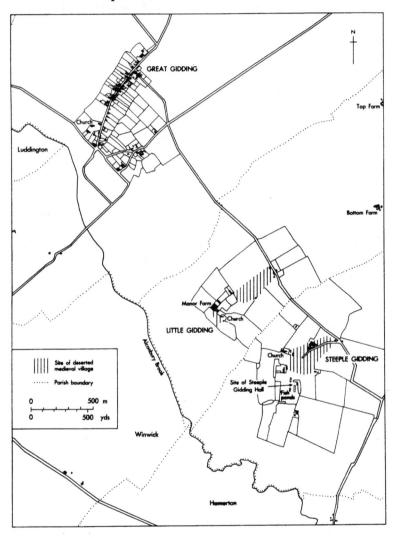

estates covering much of early medieval England and suggested that Domesday statistics of very large manors are, in fact, hiding a number of settlements under the estate name,[18] a feature first noted by Maitland. If this is so then the use of Domesday Book as a detailed guide to particular settlements may make nonsense if we assume that only one settlement was involved.

Where the Domesday Book makes the distinction in its entries between, say, Great and Little, Upper and Lower, Magna and Parva then we can be certain that separate communities had already emerged from a larger Saxon estate. This process of fission appears as a common feature in some of the county folios and yet it is little understood. The Little Domesday, covering the counties of Norfolk, Suffolk and Essex, is prolific in its listing of churches and these often help in the identification of a number of emerging villages still recorded under the same name. South Elmham (Suffolk) is simply 'Elmham' for its numerous Domesday entries (apart from the solitary reference to Elmham Saint Cross), and yet the recording of six churches enables us to identify many of the entries with the present-day villages and parishes that are separated today by the respective patron-saint name of their churches. Each of the existing churches contains some visible work of the twelfth century.[19]

The concept of the stability of the medieval village, like that of the Dark Ages, has been undermined in recent years by a number of studies that have suggested that we should be looking for a constantly changing landscape instead. One of the lessons from the thorough excavation of the late medieval deserted village of Wharram Percy (East Riding, Yorkshire) has been that it was wrong to assume that the final deserted form of the village was also that which had existed at the time of Domesday or even at its inception at some much earlier date. Since Middle-Saxon times the village had undergone at least two changes in its form (Figure 6.3) and the evidence from other excavations supports this picture of change. No community can have stood still over such a long period; the environmental conditions that impinge upon it at one time are not appropriate for another and it should be expected as a general rule rather than as the exception that a village will grow in one direction and decline in another, producing an ever-evolving form.

The evidence from Wharram Percy at first suggested a single Anglo-Saxon settlement in the Dark Ages that gradually expanded and eventually spread westwards in the twelfth century on to the higher slopes of the narrow valley in which it lay. In the following century, marked by the arrival of the Percies in 1180, a regular series of tofts was laid

Figure 6.3: Wharram Percy, Yorkshire East Riding, a deserted medieval village

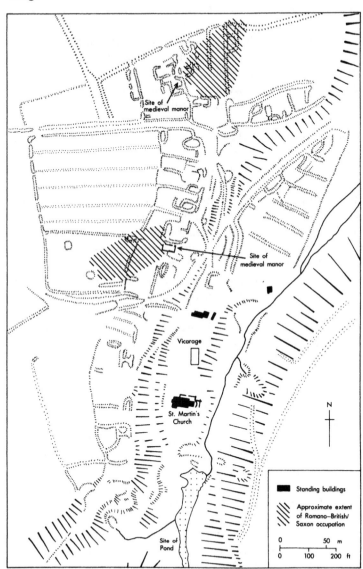

Source: Based on M.W. Beresford and J.G. Hurst, 'Wharram Percy: a Case Study in Microtopography', in P.H. Sawyer (ed.), *English Medieval Settlement* (Edward Arnold, London, 1979).

out on the former arable to the north of the village and a new manor
house was built on the northern edge, thus occupying the same position
relative to the village as its twelfth-century predecessor. But there is
some more recent evidence from Wharram Percy that suggests the
Saxon village was closely related to earlier Roman and Iron-Age communities and that, like the Northamptonshire parishes mentioned
earlier, there was more than one Saxon focus within the parish.[20] The
coalescence of these hamlets into larger medieval villages is a process
which Christopher Taylor has recognised as an important aspect of
our understanding of the evolving village form. The term 'polyfocal
settlement' was aptly applied by a group of extra-mural students who
were puzzled by the recognition of more than one focus to the early
medieval village. In the majority of cases the foci could be identified
with small manors and Taylor has suggested that many medieval villages
can be usefully explored with this concept of sub-units in mind. Many
villages cannot be easily described as a single entity, such as a 'green'
village or 'street' village, and their morphology is best explained in
terms of distinctive groups of buildings around, perhaps, a small green
or a road intersection.[21] Some settlements have a 'loop' form, their
buildings strung out along a number of twisting lanes and with particular foci identified by such names as 'Church End', 'Wood End',
'Broad Green' and so on. Cranfield in Bedfordshire is typical of a number
of villages in that county with its buildings forming a continuous line
along the lanes that link up the medieval foci of Church End, East
End, Broad Green and Gossard's Green. Such secondary place-names
are not difficult to identify from late medieval documents and it is
likely that the majority had been in existence for some considerable
time before they first appear recorded; some certainly lie silent amidst
the Domesday details. Yet the clear identification of Domesday manors
with certain individual foci is no easy task, although the details of
manorial history provided under each parish in the Victoria County
History can often offer a useful start to an investigation.

Even where individual medieval foci can be recognised in the present
landscape, their significance in the total plan of village evolution is not
easily discernible. Some may reflect a primary pattern of settlement
that has its roots back in Roman or even prehistoric times, others a
process of secondary settlement following the final clearance of woodland and waste in the late Dark Age-early Middle Age period. In central
Northamptonshire, for example, the parish of Great Doddington has
produced no fewer than six early Saxon sites and Taylor believes that
the pattern is more akin to that which we now know existed in Roman

times than to the classic open-field, nucleated village of the medieval period.[22] Elsewhere, it is well to remember that the twelfth and thirteenth centuries are noted for the frequent subinfeudation and creation of new fees which themselves generated new foci. Some of them were close enough together to create, within a very short period and under sustained population increase, the irregular straggling villages we see today. Many of the moated sites in the landscape probably owe their origin to this process, and they too form an integral part of many villages as much as they are to be found scattered on the perimeter of the parishes. Colne in Huntingdonshire has two small moated sites on the western edge of the village, separated by the ruins of its medieval church. The original Domesday manor is represented by the more westerly of the two moats, but with the creation of another medieval manor the development of the village moved eastwards and the earlier village was eventually deserted. At Icklingham (Suffolk) the churches still stand as a vivid reminder of the dual-focus origin to the village, with All Saints at the western end and St James on the eastern edge of the street village that runs parallel to the River Lark.

If coalescence of sub-units is an explanation for some of the villages of the east Midlands and eastern England it is certainly not the answer for them all. The opposite process, of dispersal from a Dark-Age centre, might explain the straggling common-edge settlements that were until recently an important feature of East Anglian villages. Dr Peter Wade-Martins's work on the central Norfolk hundred of Launditch[23] has revealed a fascinating pattern of change of village form that is not just contained within the medieval period, but which recognises again the constant adjustment of the community to changed circumstances. Through a thorough investigation of the field evidence for lost settlement patterns and the information provided by a particularly good set of early estate maps of the late sixteenth and seventeenth centuries he identified a number of Middle-Saxon nucleated settlements. The scatter of pottery from these communities around the isolated churches of the hundred confirmed the suspicion that the latter had displayed a high degree of inertia while the settlement that had formerly led to their construction had long since moved away. The evidence produced by field-walking is for a desertion in the early medieval period when the community migrated to a common-edge situation elsewhere in the parish. At Longham (Figure 6.4) and Weasenham St Peter this movement was followed by a further migration in the late medieval period to another common. Faden's detailed county map of the late eighteenth century reveals a Norfolk landscape of long irregular-shaped commons interconnected by wide drove-ways and strips of common pasture.[24] It was possible to walk through a number of parishes in the centres of the county without leaving common land. But the majority of the commons did not survive the enclosure movement of the eighteenth and nineteenth centuries. The old farms can still be recognised lying back from the new straight roads that were driven across the commons, but it was the enclosure roads that were to form the focus for new settlement and contribute to the roadside villages of today.

The reasons for the early and late medieval migrations await detailed study, although Dr Wade-Martins has already emphasised the importance of a declining amount of waste available for the pasturing of animals as the expanding medieval population used up most of the suitable land for arable. Farmers became more dependent upon the remaining areas of common pasture and it would seem logical to abandon housing upon potentially good land if a waste-edge location was available. Excavations of medieval peasant houses at Thuxton and Grenstein support the evidence of the Domesday Book, that the East Anglian peasantry were a prosperous group, small farmers rather than a landless labouring class,[25] and each would have at least one animal to graze on the common pasture of the parish. More research is needed into the significance of the woollen industry to the East Anglian countryside, for its profitable nature in the later medieval period may well have dictated a tighter control over the diminishing resource of suitable grazing land for the large flocks of sheep.

It remains to be seen whether the model of migration on to the edge of commons has a wider application than the clay belt and sheep–corn husbandry of East Anglia, for such common-edge villages are to be found over a much wider area of southern and eastern England. The green villages of northern England are, however, rather different and their examination introduces an important element, the planning or replanning of a village on simple but rigid lines.

The morphology of the villages of Durham and Yorkshire is more readily categorised than most. The late Professor Thorpe was able to place many Durham forms into 'broad-green', 'street-green' and 'tri-angular-green' from the nineteenth-century Tithe and Ordnance Survey maps[26] and with the inclusion of the two-row or street village this accounts for the majority of the rural settlements. But an error was made in assuming that such nineteenth-century forms had remained basically unchanged since the settlements had acquired an Anglo-Saxon or Scandinavian place-name. A millennium or more of turbulent history of a border county negates such an assumption. Yet it appears

Figure 6.4: Longham Parish, west Norfolk, settlement migration

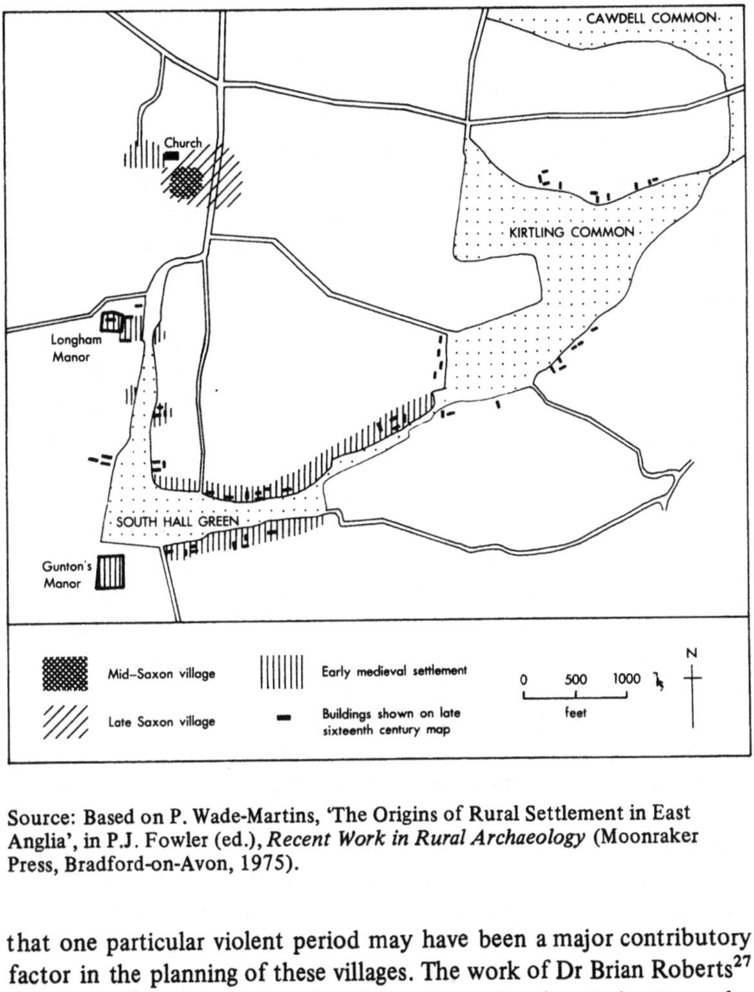

Source: Based on P. Wade-Martins, 'The Origins of Rural Settlement in East Anglia', in P.J. Fowler (ed.), *Recent Work in Rural Archaeology* (Moonraker Press, Bradford-on-Avon, 1975).

that one particular violent period may have been a major contributory factor in the planning of these villages. The work of Dr Brian Roberts[27] and others[28] has suggested that the early medieval period witnessed a major rebuilding of villages in northern England with tofts being laid along regular lines and using standardised measures. Manorial documents attest to the existence of two-row, and in some cases three- and four-row settlements on the manorial estates of the bishops of Durham, new settlements laid out using the rod or pole as the basic unit of measure. The length of the medieval rod varied from place to place and did not always conform to the standard length of 16½ feet; on the bishops' estates 20 feet was the norm. In some cases the documents provide a clue to the location of the rows by the inclusion of a compass point, hence 'West Row' and 'East Row'. Through a combination of documentary research and an examination of existing toft boundaries on the ground Dr Roberts suggests that many Durham villages were replanned in the twelfth and thirteenth centuries as a necessity, follow-ing the devastation of the northern borders that has been alluded to earlier. In some areas villages would have been rebuilt, but in others new sites may well have been chosen and the planned village appears to have grown at the expense of smaller hamlets located nearby. Ay-cliffe, a large multiple-row green village, was listed in a gilly-corn schedule along with the villages of Newton Ketton, Chilton, Woodham and the hamlet of Woodhouses. Gilly-corn rents were paid by all bond and free tenants of the monks of Durham as a contribution to the provi-sion of alms and were levied on lands brought into production before 1200. Significantly, Newton Ketton was already mentioned as early as 1235 as having 'formerly' sixteen bondages or tenants and the settle-ment had completely disappeared by the early fifteenth century. Woodham suffered a similar fate, Newhouses can no longer be traced in the landscape and although Chilton still survives aerial photographs have revealed considerable evidence of shrinkage. It seems likely that the new village of Aycliffe had absorbed some of the population from the other communities. In the same way the two-row green settlements of Middridge, Cowpen Bewley and Bolam can show evidence for an 'Old Town' site lying amidst the fields.[29]

June Sheppard's recent analysis of the regular village forms of York-shire has shown how they are located principally in the lowland areas.[30] As it is unlikely that such areas were not supporting a high density of population in the early eleventh century, then the regular-planned settlements cannot be regarded as a feature of new settlement upon waste or marginal land but rather as a reorganisation of existing settle-ments. Following the devastation of William's campaign of 1069–70, she suggests a two-stage recolonisation of the newly created waste. Where survivors could be regrouped within a 'manorial' type of com-munity then villages might well be rebuilt upon their old lines. But in other circumstances it seems that a free rent-paying community was attracted on to a waste vill where a regular-form village was laid out. A high correlation can be found between regular-planned villages and those vills which did not possess any demesne in 1086; such vills were also more likely to be held by one of Yorkshire's 26 tenants-in-chief (those who held land direct from the king) than subinfeudated. It appears that the honorial lords were operating a policy of encouragement to free tenants in order to recolonise quickly their most devastated vills, hence the granting of a rent payment rather than more onerous bondage obligations. The laying out of a regular plan for the village would have simplified the description of the holdings and made equitable the share of rentals and dues of the community.

The simplification of landholdings and dues was not just confined to the form of the regular village, for in a number of cases the same sort of regularity can be seen applied to the layout of the open fields. Some vills, as at Wheldrake,[31] had their arable originally contained within the long tofts that ran back from the village street as part of an infield-outfield arrangement that probably characterised much of Dark-Age agriculture in northern England. More widespread was the practice of sun-division or *solskifte*, whereby the arrangement of the tofts in the village was to determine the position of the holdings in the open fields. Each toft was noted according to its location in a clockwise sequence and the individual strips in the furlongs followed the same sequence, those strips in the south and east of the furlong said to lie 'next to the sun' and those in the north or west 'towards the shadow'. This close relationship between village and field can only have occurred where the two were planned together. Dr S. Goransson has demonstrated how the system can be widely recognised in England, but is particularly concentrated in the north-east.[32]

A number of problems remain over the regular-planned villages of northern England. Not all villages fit neatly into the categories of planned and unplanned, and the north-east has a number of examples where a regular row of tofts has been added to a more irregular form; the thirteenth-century replanning at Wharram Percy illustrates this well and emphasises the danger of assuming that all regular plans in the north are a consequence of the late eleventh-century campaign. Pamela Allerston's research into the village plans of the Vale of Pickering has also suggested that the transformation to regular village may not have been as rapid as others have implied, for the creation of small hamlets, each with its own infield, may have preceded the full-blown villages of the thirteenth and fourteenth centuries. Appleton-le-Moors (Plate 6), where a classic two-row village can now be seen, developed from an amalgamation of the earlier Appleton and the neighbouring vill of Balskerby by the thirteenth century. The site of the latter remains uncertain, although examination of the nineteenth-century tithe documents and early Ordnance Survey maps show the names of 'Bawsby' or 'Boastby' about one kilometre north-east of the present

Plate 6: Appleton-le-Moors, a two-row village on the southern edge of the North York Moors

village.[33] Such late amalgamations would suggest that a different process was at work here, a social and tenurial revolution forced upom communities through the pressure of population upon meagre and finite resources. It has been argued elsewhere that the tightening of regulations for the operation of common open-field systems at the same time was for a similar reason.[34]

It is clear from these detailed studies that the examination and interpretation of medieval village forms has reached a stage of complication. The earlier, simplistic view of organic growth from an 'original' Anglo-Saxon or Scandinavian core is no longer tenable over many parts of the country. However, what is still lacking is an overall new approach to the study of the medieval village that does not rely heavily upon the availability or uniqueness of certain early medieval documents. It is medieval archaeology that holds some of the keys, but the cold facts of medieval artefacts cannot in themselves 'prove' migration from one part of a parish to another or allow us to see the reasoning of the individual lord or the community in their decision to replan the village layout. In the same way the primary–secondary paradigm that has characterised many settlement studies and has recognised the many hamlets and isolated farms in the landscape as a colonisation from the older villages needs careful reconsideration. Some writers now see the hamlets in some areas as reflecting a much older pattern of settlement that predates or is at least on a parallel with the so-called primary settlements.[35] It seems likely that we will serve no useful purpose in throwing out the old paradigm and replacing it with a new one, for the complexities of the English settlement landscape are wide enough to accommodate both old and new and there can be little doubt that the basic ancient pattern, however we interpret that, was repeatedly supplemented and diminished by changing demands and responses in the delicate balance of population to resources.

Whether we recognise the early medieval period as the main period of settlement expansion of pre-industrial times or, as others now see it, as the culmination of a far longer period of sustained colonising activity, there is no doubt that the fourteenth and fifteenth centuries were a time of retrenchment and population decline. The Black Death of 1348–50 has always been a popular target for the blame, but there is evidence for the abandonment of marginal land and some communities in the half-century before the plague had reached these shores. The debate of recent years has centred upon the archaeologists' claim of a significant climatic deterioration at the end of the thirteenth century as a major factor in the decline, whereas economic historians have popularly favoured the 'Postan Thesis' with its Malthusian explanation of a sharply rising population eventually overstepping the mark and exceeding the available resources.[36] There is evidence to suggest that both explanations may be valid in the desertions of the early fourteenth century. Hound Tor village has recently been revealed from the cover of bracken that had hidden it since its desertion in the early Middle Ages. Its position above a site at a thousand feet on the eastern slopes of Dartmoor can only be regarded as marginal, even when the settlement was established between AD 700 and 800, and with the onset of a wetter regime in the thirteenth century the cultivation of cereals was no longer viable. A succession of turf-walled longhouses had been replaced in the last century of the community's life by more substantial stone-walled structures. The lower courses of four long-houses, four other houses, a shed and three corn-drying barns can now be recognised in the landscape. Apart from minor changes in the alignment of the houses, the size and form of the settlement had altered little in its entire history. The corn-drying facilities attest to the fact that what had begun as a minor inconvenience to the community when wet, cool summers required the harvest to be made before the grain was fully ripe or while it was wet had eventually turned into a major disaster. The conversion of some houses into barns in the dying years of the hamlet shows that the process was a gradual one. Yet it is almost certain that it was over, as it was for other communities in similar locations, before the first outbreak of bubonic plague.[37]

The deterioration of the climate did not just affect upland villages, for those on the heavy clays of the lowlands could be at a similar disadvantage. Even a slight increase in rainfall could render many fields unworkable, and this may have been an important factor in the demise of Goltho in Lincolnshire, lying as it does amid a great cluster of medieval deserted sites.[38] But elsewhere it could be argued that soil exhaustion was a more important consideration. The returns made for the taxation of the *Nonarum Inquisitiones* of 1342 contain a number of pleas from communities who argued that they were impoverished through the poorness of their soil.[39] Vills on the light sandy soils figure more prominently, as on the Lower Greensands in Bedfordshire, where the villagers of Potton complained 'the soil is sandy there in a dry year and little value'.[40] In Norfolk 34 of the Domesday place-names do not appear in the *Nomina Villarum* of 1316, and the majority may be traced to the poor sand region of Breckland and west Norfolk.[41]

Although a number of the medieval desertions can be placed in the ealy centuries of the period the majority belong to the main phase of 1450–1520, when landlords took the final step of removing the few remaining tenants of an impoverished community so as to convert the land to sheep pasture. The medieval documents of taxation, particularly the Lay Subsidy Returns for 1334 and 1428 and the Poll Tax of 1377, enable us to gauge the size and prosperity of many villages and it is clear that most of them had significantly smaller populations in the fifteenth century than they had enjoyed two centuries earlier. For the landlord this had meant declining rent rolls and an increase in the cost of hired labour; when the price of wool rose relative to grain it was tempting for many of them to turn a partly occupied vill and under-used fields to better profit under sheep.

Professor Beresford, in his excellent study of medieval deserted villages,[42] has cogently argued for the importance of the individual landlord in determining the pattern of desertion. For while the Midland counties, stretching from the Dorset coast to Yorkshire, contain the largest number of desertions because of the prevailing tenurial and environmental conditions to be found in them, the individual desertion can only be properly understood through the decisions and whims of the landowner. Where the demographic changes of a local landed family brought them ill-luck in the absence of a male heir, then a long family association with the manor might be broken and the incoming family less sympathetic to the problems created through enclosures. Estates with absentee landlords were also more prone to depopulation. Wormleighton in Warwickshire suffered from both of these features, for the manor had fallen upon mixed fortune with the death of Sir John Peche in 1386, leaving but a widow and two daughters to inherit, and by the close of the following century the manor house was in disrepair as it no longer acted as a principal residence. In 1499 it did not prove difficult for William Cope to buy out other minor landowners and remove the occupants of the remaining twelve messauges and three cottages, sixty persons in all, and to convert the parish to sheep pasture.[43]

Much of the deserted village of Wormleighton can still be seen in the landscape and aerial photographs have recognised that the existing settlement that clusters around the early Tudor house the Spencers built lies partly over other house plots (Plate 7). The low regular-shaped humps and narrow hollow ways mark the position of individual house crofts and the lanes that formerly ran through the settlement,

Plate 7: The deserted medieval village of Wormleighton, Warwickshire

and these are a common feature of deserted villages where the tradition of pasture has continued since their enclosure. In those counties where intensive, deep ploughing is now practised the recognition of desertions is often only made clear through the use of aerial photography and the activities of the Air Photographic Unit of Cambridge University under the auspices of Professor J.K. St Joseph have done more than any other agency to uncover lost sites in this manner. Ploughed-out sites are certainly more difficult to interpret for the spreading action of repeated ploughing produces a 'smudged' effect on the rubble footings and collapsed walls of wattle and daub that characterised most peasant houses. Similarly, the blackening of soil that middens and hearths produce with prolonged use are difficult to pin-point as the soil-mark is spread over a considerable distance.

The investigation of any deserted site is not only interesting for what it shows us of village streets and peasant housing, but also for revealing the location of the more important dwellings of the manorial community. In some of the small poverty-stricken manors we cannot expect the manor house to be any larger than most of the other properties, but the larger vills usually have a much bigger regular mound to identify it. In a number of places, especially those villages on heavy clay soils, the manor house had a further distinguishing feature, a rectangular moat. At Great Streeton in Leicestershire the moated manor-house site is clearly identifiable on the southern edge of the deserted village, alongside a smaller depression that marks the site of the manorial fish-pond.[44]

The great majority of deserted medieval villages are most readily identified through documents or from landscape evidence. But a number of lost villages have also been located through the careful examination of private estate maps of the late sixteenth and seventeenth centuries. Only the village of Boarstall in Buckinghamshire can boast of a pre-desertion map that shows its medieval form; the drawing itself is crude and not to scale, and was only intended as an illustration to a manuscript. True maps, those drawn up by surveyors, were not a feature of the late fifteenth century and do not appear in any number until a hundred years or more after the majority of desertions were over. However, they can help in the identification of the site, for the surveyors sometimes marked on 'Old Town', or the new fields themselves had taken on indicative names; Chalford in Oxfordshire can be identified from a mid-eighteenth-century map as 'The Towns Coppice', 'The Towns' and 'The Towns Piece'.[45] Other names may indicate recent enclosure and their great size makes us suspicious that they contain not only the former arable but also the remains of the village. The fine estate map of Sawtry in Huntingdonshire in 1612 records 'The Greate Pasture' in the south-west of the parish, a single field covering a massive 508 acres. It is unlikely that this was part of Sawtry's open fields, for they are still shown, covering almost 1800 acres. The answer perhaps lies in the neighbouring parish of Coppingford, for the tell-tale humps and hollows of desertion lie within a few yards of the parish boundary and 'The Greate Pasture'.[46]

Seen from the air the form of Coppingford (Huntingdonshire) is identified with a large triangular green. There is a need for great care, however, in assuming that the green form was either the final shape of the village or the focus of its morphology for any length of time. In extreme cases it could be argued that the village had never had a triangular green, for what we see from the air or on the ground is the final picture of composite form, a picture built up by successive layers of changing layout and rebuilding. Each building plot and each lane has left an indelible mark in the landscape, no matter how faint some of those traces are, and those changes in layout that have succeeded them will have been unable to erase them. Only the systematic excavation of a larger number of deserted villages can provide the clear exposition of medieval village form that we seek, and the number so excavated is pitifully small. Given the very considerable quantities of capital and labour required for such operations the situation is regrettable but perfectly understandable.

The excavations that have been carried out have already greatly expanded our knowledge of the nature of building form and the structure of medieval houses. Wharram Percy[47] and Goltho,[48] for example, have revealed the clear shift from timber and clay or turf-walled structures to a much greater use of stone in the early medieval period, in those areas where stone was readily available. If this was a reflection of a shortage of good building timber then it is perhaps not surprising to see something of a resurgence of timber framing in those same areas at the close of our period, when the desertion of villages had allowed the rejuvenation of woodland on abandoned fields and pastures. The rate at which buildings were renewed has also aroused considerable interest. Most peasant hovels were built of such flimsy materials that they required rebuilding every 25 or 30 years. What is more surprising is that rebuilding often involved a change in the long-axis direction of the building. Examination of the two-row village of Holworth in Dorset suggested that this may have been a way of accommodating a shift in the direction of the prevailing wind, which in itself would have been an important element in any amelioration or deterioration of climate.[49] These long rectangular structures usually had an entrance in the middle of the long sides, and the gable end would therefore face into the worst of the weather. At Wharram Percy, however, some of the realignments were so drastic that they suggest that a change of ownership or generation may have been more important than a subtle adjustment to longterm climatic fluctuations. The frequency of rebuilding may have been conditioned by a custom of providing a new house when son succeeded father to the tenancy as much as by the flimsy nature of the materials employed.

The Medieval Village Research Group now continues the pioneer work of Beresford and Hurst in the identification of lost sites and has also broadened its scope in the analysis of the medieval village, deserted or otherwise. One important feature has been the recognition of shrinkage of villages as equally worthy of our attention, for most villages that are as yet unaffected by recent growth of commuter homes exhibit some degree of shrinkage, although not all of it can be ascribed to the medieval period. We have also come to the realisation that the study of deserted villages is not in itself a satisfactory way of understanding all medieval villages. The changing form of the settlement indeed suggests caution in seeking standardised forms and the small nature of many late medieval vills that suffered depopulation may not have much to tell us of the larger ones that survived. It is doubtful if smallness implied a certain settlement form, although there is a tendency for deserted villages to have a very simple form, the two-row or street village being the most common. It remains questionable whether deserted villages can tell us any more about similar-sized communities that survived than about those that were larger, but we have already learnt a great deal about the peasant farmer and the small seigneurial family. Their livelihoods were intricately bound together and one suspects from the villages excavated that they were little divorced in terms of wealth and possessions.

Towns

The function of the medieval town was more sharply differentiated from that of the village than is the case today. Apart from those whose livelihood depended upon the patronage of a castle or great monastery the majority of the medieval towns were concerned with trade and with providing a wide range of manufactured goods for town and country.

Many enjoyed a high degree of independence and were self-governing, although this feature was more pronounced in those towns where charters were already of ancient standing by the twelfth century.

It has been a common practice in recent years to discuss medieval towns in categories of planned and unplanned (organic) growth. The twelfth and thirteenth centuries witnessed a massive growth in the number of towns, responding to the sustained increase in agricultural activity of the period, and these new towns have often been erroneously labelled 'plantation' and 'planned'. But of the hundreds of new towns to appear, such descriptions can either fit a few or all of them. Many new towns can rightly be regarded as plantations, where they were deliberate new starts on virgin soil, often the poorer soils of a parish or estate; however, a greater number are involved in those that grew out of existing villages, either through a raising of a community to borough status or through the addition of a borough sector. As to the planned aspect of medieval towns, Aston and Bond have rightly argued that there are very few towns where some planned element cannot be recognised,[50] for most of them were under the control of an individual or small ruling elite. Their decisions to restrict development in one part of the town or to encourage it in another were invariably of considerable consequence for the town's morphology. Contrary to popular belief only a minority of plantation towns exhibit a regular plan on a grid, a 'once-off' establishment that fixed not only the core of the new town around its market-place, but also provided its bounds before it was even occupied. Such towns displayed the maximum confidence that all their plots would be taken up; they belong to the euphoria of the early twelfth century when the town-founders were at their most optimistic. Professor Beresford's lists of new towns include only 26 grid-iron plans, including New Sarum, New Winchelsea and New Buckenham.[51]

The planning of towns is not a medieval revival of a lost Roman art, for there is now sufficient evidence to recognise Dark-Age origins for many towns and certainly an element of planned growth that pre-dates the Conquest. The street pattern of Winchester may well poorly disguise a former Roman pattern,[52] while a number of the Roman towns provided a fixed point to later patterns in those streets that focused upon a gateway through the city wall. In Bedford the regular grid of streets that form the town centre seems likely to have been established when the Saxon burh was created in the ninth century.[53] The walled towns of Wallingford and Wareham exhibit clearly the features of the late Saxon planned town.

Royal foundations and those of the monasteries played an important part in the wave of new-town creations of the late eleventh century. Some of the towns were secondary elements to military considerations but this feature was less well developed compared to the castle towns of Wales founded by Edward I. Before he succeeded to the English throne Edward had received a good grounding in new-town foundation through his experience with the bastide towns of Gascony. Apart from his successful military foundations in Wales, he was instrumental in the creation of New Winchelsea, Kingston upon Hull and Berwick-upon-Tweed.

The Domesday Book is unfortunately silent or restrictive in what it tells us of the immediate post-Conquest situation in the English towns, although it does present a picture of temporary decline in some of the older-established boroughs. Some towns witnessed the intrusion into their plan of a new or greatly enlarged castle, as part of William's efforts to dominate the shires. In Huntingdon the expanded castle entailed the demolition of houses, including a town house of the Bishop of Lincoln, and it appears likely that a similar fate had affected part of Bedford.

For a few towns the Domesday Book is something more than just the listing of a market and the mention of burgesses, and none is better served than the new town of Bury St Edmunds, where a town had arisen on the former fields of the settlement since the Conquest. The shrine of St Edmund, patron saint of England before his usurpation later in the medieval period by the foreign import of St George, was to make the Benedictine monastery at Bury wealthy and famous. As a centre for pilgrimage its potential for a profitable market had been quickly realised by the Norman abbot who laid out the ambitious grid-iron town between 1066 and 1086. After describing the pre-Conquest manor the Book says:

> all this account refers to a town as if it were still as in the reign of King Edward. But the town is now contained in a greater circuit including land which was then being ploughed and sown.[54]

In all, 342 houses had been erected, a great open space — Angel Hill (Plate 8) — laid out before the abbot's gate, and markets provided for butter, horses, beasts and corn.

The commercial value of pilgrims to a saintly shrine is not lost to entrepreneurs even today, and Bury St Edmunds is no solitary example from the Middle Ages. St Neots and St Ives (Huntingdonshire) both

Plate 8: Bury St Edmunds, Suffolk: the Abbey precinct is in the top right and immediately to the left of it is Angel Hill

developed as planned towns at the instigation of their monasteries, as did those of St Albans, Newton Abbot and Battle. The market-place of Wymondham in Norfolk stands half a mile away from the small rural settlement of 1086, but outside the gates of the Priory, founded in 1107. St Ives and St Neots had the added advantage of a riverside location that enabled pilgrims, and traders, to come from a very wide area.[55] The annual fair of St Ives was a grand affair, attracting custom from much of eastern England and the Low Countries until eclipsed by the nearby Stourbridge Fair at Cambridge. The town's elongated market-place runs parallel to the Great Ouse, its fifteenth-century bridge feeding pilgrims straight into its centre and with narrow passageways running between market and riverside. These 'Waits' of the town are still not perfectly understood, but they probably filled the same function as those still to be seen in Great Yarmouth, allowing easy access for men and small carts between river and market.

The presence of water, either as a river or along the coast, acted as a natural break-point in many travellers' journeys and hence proved particularly attractive for new-town foundation. Those established on the coast were often the most successful and the new towns of King's Lynn, Newcastle upon Tyne, Boston and Kingston upon Hull figure prominently in the Lay Subsidy Returns for 1334 and the 1377 Poll Tax, having large numbers of taxpayers. Those at riverside crossings were also assured, at least in prosperous, well-populated areas, of a good start. Stratford-upon-Avon, Hungerford and Chelmsford all had no problem in filling up their burgage plots. Some towns had further benefit from a border situation, where the uncertainty over control meant an avoidance of taxes, dues and other customs of the adjoining manors. However, it should not be assumed that such towns deliberately chose those sites for that reason. Where a market was established astride a major highway it meant that a border situation was unavoidable if the road had been previously used to mark, say, a county boundary and had fallen into disuse temporarily as a highway. Royston straddled the boundary between Hertfordshire and Cambridgeshire, its triangular market-place situated across the old Roman road of Ermine Street. In the Dark Ages the former route to Cambridge had been abandoned in preference for one to the east, through the villages of Barley and Barkway, but its medieval revival had encouraged the foundation of Royston. Newmarket lies in a similar border situation, its long narrow street following the ancient Icknield Way and with the houses on one side in Suffolk, those on the other in Cambridgeshire. Buntingford in Hertfordshire was in an even more confused state for its establishment along the main road of Ermine Street split its control among four rural parishes, two of which were later to be deserted.[56]

By no means all the new plantation towns or the so-called organic boroughs were successful. The only medieval foundation after 1348 was Queenborough in north Kent, intended as a naval base for the Hundred Years War, and its single main street is a vivid reminder of the fact that late foundations stood much less chance of success.[57] Any town coming late into the race, when agricultural activity was already in decline and a century of plague lay ahead, needed exceptionally good fortune. Even royal patronage was no guarantee. Edward I established Newton on the shores of Poole Harbour in 1286 as a port from which valuable Purbeck marble could be exported, but by 1558 only one cottage remained (Newton Cottage) and it appears that the town never got underway.[58] On the Isle of Wight the similarly named Newton, founded in 1256 by the bishops of Winchester, experienced a slower decay, but had gone by the eighteenth century. Others suffered more violent fates, for the natural elements were no respecter of early or late foundations; Ravenscrodd had been established on the newly cast-up sands of Spurn Head between 1240–1250, but it was a venture that allowed nothing for the vagaries of the sea. Two-thirds of the streets had been washed away in 1346 and within twenty years the borough ceased to exist.[59] The late medieval fate of the larger town of Dunwich on the unstable low cliffs of Suffolk is too well known to repeat here in detail.[60]

Even if a town thrived during the medieval period it did not necessarily follow the same prosperous route afterwards. New Buckenham in Norfolk is now no more than a small village, yet its grid-iron plan had been successful after its mid-twelfth-century foundation. When the medieval church was constructed to replace the chapel that had formerly served both town and castle, it had to be placed outside the grid's chequers, presumably because they were already occupied.

For the non-plantation towns the medieval period was also one of considerable change. The granting of market charters to raise villages to town status was a widespread practice of the earlier centuries and others had major new additions and a market charter either granted or transferred from the old settlement. Bicester in Oxfordshire has a distinctive planned element in its spacious High Street (Figure 6.5), but its road pattern and wide burgage plots lie at right angles to the older sector of Kings End, with its twisting streets and parish church. Professor Beresford lists St Ives (Huntingdonshire) as a plantation town, but in reality it too has two sectors, the parish church on the western end

Figure 6.5. Bicester, Oxfordshire: a medieval new town attached to an earlier centre

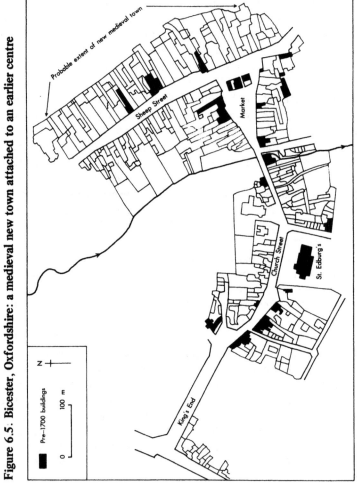

Source: Based on data in K. Rodwell, *Historic Towns of Oxfordshire* (OUP, Oxford, 1975).

of the town reminding us of the Domesday hamlet of Slepe that pre-ceded the town. However, the granting of a market was of itself only part of acquiring urban status. In the prosperous Essex countryside, market charters were obtained from as far apart as Castle Hedingham, Burnham-on-Crouch, Hatfield Broad Oak and Newport but none of their names survives in a list of markets in 1575.[61] Yet in each of those villages it is possible to recognise the effect that the market has made, widening the street to accommodate it. The morphology of the market-place dominated the form of most towns, for while in the plantation grid town it occupied but one chequer, in others the town was moulded around it. Where the town lacked defences a planned layout of burgage plots was often restricted to the market frontage, and subsequent back-street development was of a more haphazard nature. These marginal locations were clearly less desirable and the clamour for a market-place position probably accounts for the widespread encroachment on to the open market by the end of our period. Temporary daytime stalls had gradually been replaced by permanent structures and the original market-place lost in a dense concentration of buildings separated by narrow alleys. John Speed's fine town plans of the early seventeenth century, included as insets to his county maps, reveal how far the county towns had infilled their centres in this way. The markets of Hereford, Huntingdon and Ely are barely recognisable.

Unlike the older-established county towns the new medieval planta-tions were rather more generous in their allocation of burgage plots and the width of their streets. The Bishop of Worcester allowed 50 feet for his roads at Stratford-upon-Avon and the market street was 90 feet across. The quarter-acre burgage plots, which were related to former field widths, were 200 feet in depth and gave a frontage of 60 feet; those in the new borough of Salisbury were only slightly narrower at 50 feet.[62] But at Alnwick in Northumberland, where there is still some debate over a pre- or post-Norman foundation for the town, Professor Conzen's detailed study revealed almost half the medieval frontages were measured to 28 feet or recognised fractions of the same, while many others related to a 32 feet width. As many of the medieval houses faced their long gable onto the street, and houses were traditionally of one or two bays, with the standard bay of 14-16 feet, then it is clear that the structural arrangements of the house had a strong influence on burgage width.[63] Most towns contain some survivals of their original burgage plots, although their bottom ends have sometimes been sacri-ficed to new relief road schemes or car parks. At Olney in Buckingham-shire, where a medieval borough was added to an existing village, the back lanes of West and East Street still preserve the line of the back end of the burgage plots and many of the original boundaries survive. Glastonbury (Somerset) has truncated the plots on the north side of the High Street for a car park but stone setts mark the extent of one, at the rear of the fifteenth-century Tribunal House.[64] The building is now a museum, but was formerly the court house for the abbot of the wealthiest monastery in southern England, at whose gates the small town had developed.

Although there are references to plot subdivision as early as the thirteenth century, when half- and even quarter-burgages were recorded, the excavation of some town sites has suggested that many boundaries are of considerable antiquity. In Westwick Street, Norwich, property boundaries remained unchanged from the twelfth until the eighteenth century, while excavations in nearby Oak Street have revealed a similar pattern over a shorter time period. Even when the town experienced considerable population expansion in the sixteenth and seventeenth centuries, this was carried out within the medieval boundaries, by infilling and adding a further storey.[65] Where sub-division did occur, it forced a change in building alignment, the long axis turning at right angles to the street. Stamford and Stratford-upon-Avon both exhibit this feature clearly.

The examination of medieval buildings that survive in our towns is still in its infancy.[66] The recording of buildings is spasmodic and often uncoordinated, and there is no detailed survey of the town houses of England apart from Professor Pevsner's necessarily brief comments in his county series, *The Buildings of England*, where some of the evidence is second-hand or based upon external evidence alone. More late medieval, and some much earlier structures, survive behind Georgian and Victorian facades than most people realise. The examination of many town houses is often best conducted from the rear of the properties for that very reason, although dating is usually only accurate if a full internal survey can be carried out. The Royal Commission on Historical Monuments is slowly making good the deficiency, but many of its earlier volumes deal more adequately with the countryside than the town or do not differentiate between buildings of pre-1700 and the late medieval period. The newer volumes devoted to Stamford, Salisbury and York show what can and should be achieved.[67]

As a habitable structure the medieval town house has a better record of survival than its rural counterpart, although in both cases it is only the residences of the rich that have had any chance of surviving. The so-called 'Jews' and 'Normans' houses of York, Lincoln and Southampton, together with Moyses' Hall at Bury St Edmunds, are all good examples of mid and late twelfth-century domestic architecture but such stone-built houses cannot have been common in the medieval town. In spite of their names there is no record that they were ever owned or dwelt in by Jews but it has long been a popular fiction that only a money-lender could afford such a house. Even in areas where stone was readily available timber-framing was the norm for town building.[68] Stamford's rebuilding in the seventeenth century has obscured many of the older structures of the town, but even the new construction relied upon a timber frame for the new stone-clad and stone-roofed houses.

Not all the early medieval structures have survived so well as those mentioned above and the demolition of 28–32 Queen Street, King's Lynn in 1976 is a salutary warning of how an unrecorded building can escape the regulations controlling the demolition of historic buildings. What to the outsider and to preservation societies had appeared to be a row of uninteresting nineteenth-century cottages proved to contain a stone house of *c.* 1200, the second-oldest structure known for the town.[69]

Of later medieval structures we now have a clearer picture, either from standing examples or from excavation. Some of the smaller towns still possess a high proportion of houses from this period. Yet it is those very towns where archaeological and architectural vigilance can be at its weakest and where town redevelopment schemes can do an immense amount of damage to our heritage.[70] The mixture of architectural styles in Ware (Hertfordshire) contains many timber-framed structures of the late fifteenth–sixteenth century, perhaps reflecting a period of prosperity for the maltsters who provided most of the town's employment. Place House in Bluecoat Yard, just off the main street is of national importance, for its seventeenth-century facade hides an early fourteenth-century aisled hall-house.[71]

The medieval church has survived as the most visible remnant of the medieval town, although other ecclesiastical buildings (monasteries, chantry chapels and hospitals) have fared badly. The dissolution of the monasteries in the sixteenth century was often followed by widespread pillaging of monastic sites to provide building stone for the rest of the town; parts of the magnificent abbey church of Bury St Edmunds can be recognised in tiny pieces of carved stonework scattered throughout the town's buildings. Almshouses or hospitals stood a better chance of being undisturbed, although those of a secular nature are now the most conspicuous. Lord Leycester's Hospital in Warwick was one of the few timber-framed late medieval buildings to survive the holocaust of a fire that destroyed much of the town centre in 1694. Formerly a guild house, it had been tranformed into almshouses in the late sixteenth century. Browne's Hospital in Stamford is another well-known example, this time in stone; founded by William Browne in the reign of Henry VII it still continues its function as an almshouse.

The high density of medieval churches in the major towns of London, York, Bristol and Norwich (Figure 6.6) is a noteworthy feature of the townscape, although many of the actual structures have not survived to the present day. Norwich now boasts the highest number of medieval churches of any English town, although many no longer keep their ecclesiastical purpose, being ingeniously converted to a church museum, Scout headquarters and drama centre. In the old established towns medieval churches would have been encountered every few streets or so and this does not take into account other ecclesiastical buildings that were not part of the parish system. Although some of the churches can be explained in terms of a guild association, the majority of them reflect the very high densities of population to be found in most quarters of the medieval town. In contrast, the new plantation towns are often characterised by large parishes and a solitary medieval church. At their foundation many of them were initially only endowed with a chapel of ease, the rights to the important offices of baptism, marriage and burial being reserved by the existing church out of whose parish the town was carved. This was the case at Market Harborough (Leicestershire), built within the parish of Great Bowden, a mile or so across the fields from the village, in the middle of the twelfth century. It was not until a century later that the new town was granted its own church and even then Great Bowden retained the rights to burial, a valuable source of revenue to medieval rectors. Hence the tall spired church of St Dionysius has to this day no churchyard or burial ground.[72] This is a feature to find repeated through many of the plantation towns.

The church–market-place relationship in many of the new towns was also of considerable importance. Boston's famous 'Stump' dominates its massive market-place, acting not only as a beacon for vessels across the treacherous sands of the Wash but as a reassurance to the town's burgesses of the success of their new venture. For those towns where pressure of population saw the creation of a second parish (or, as at Heden in Yorkshire East Riding, and New Salisbury, three) within the medieval period the new church did not carry the same amount of civic pride and it is the market church that stands supreme. King's Lynn was in a rather unusual situation, for its second church was

Figure 6.6: The late medieval morphology of Norwich

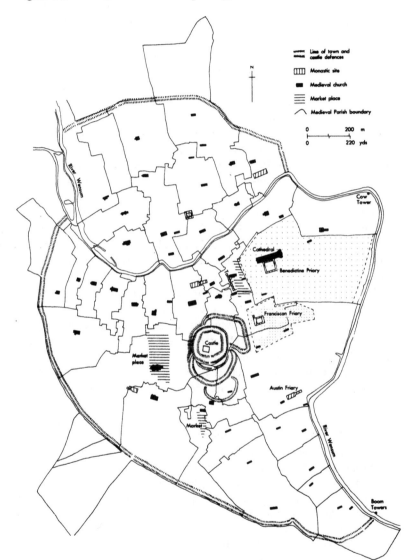

Source: Based on data in *The Atlas of Historic Towns, Vol. 2* (The Scolar Press, London, 1975).

also placed alongside a market, that which accompanied a northward extension of the town in the middle of the twelfth century. The church was, nevertheless, a dependent chapel and did not achieve full independence until the nineteenth century.

The majority of the larger medieval towns and many of the smaller ones were provided with defences.[73] With the unsettled political condition of the country throughout the early centuries of the period few towns took the chance of leaving themselves entirely unprotected. A number of the shire towns do not appear to have built formal defences, but towns such as Buckingham and Hertford were provided with powerful castles. At Bedford the strengthening of the Norman castle in its position within the town centre may have been sufficient reason for the ditches and banks of the pre-Conquest town to be ignored. Other towns, such as London, Colchester, Exeter and Chichester were able to make use of the line and materials of the previous Roman town, while York, Chester and Hereford made partial use of them. At St Albans, Ipswich, Cambridge and Lichfield earthen ramparts and an outer ditch were either considered sufficient or were all that the town could afford. Murage grants, which record the tolls levied for the upkeep of walls, show a very considerable amount of activity down to the middle of the fourteenth century. After that the threat of disturbance did not necessarily decline, but the improved methods of attack and siege warfare rendered effective defences too expensive for most towns.[74] What is noticeable is the substantial anti-cleric feeling that developed in the towns such that monasteries and cathedral closes began to protect themselves more effectively from the populace. Thus the Bishop of Bath and Wells built a substantial moat around his palace in the early fourteenth century, crossed through a defended gateway that was added in the following century. The massive gateways that survive at Bury St Edmunds and St Albans are a reflection of the fear that beset fourteenth-century abbots.

The cramped nature of the medieval town within the walls was accentuated by the large amounts of space devoted to non-residential purposes. At Exeter the cathedral close occupied almost a third of the town, with the castle taking another small area. Even extensive Norwich, its walled area five times that of Exeter, lost 16–18 acres in its very core to the huge Norman castle keep and its outworks (see Figure 6.6). The cathedral close occupied a much larger area. At Lincoln, if we exclude the large walled suburb of Newport, the castle and cathedral together controlled over a third of the old Roman town area even though much of the cathedral close lay outside. With the expansion of some towns in the early medieval period it is perhaps not surprising that extensive suburbs developed beyond the security of the walls.[75] At Bristol, Lincoln and York the new suburbs were provided with defence through an extension of the walls, while others stretched their jurisdiction over them but did not see any purpose in providing a wall, as the City of London did in 1222 when its boundary was moved westwards and northwards. Many Continental towns have a series of walls that mark successive phases of growth, some of them covering the post-medieval period. From what we know of the squalid nature of most English medieval suburbs, for example from late medieval and sixteenth-century taxation lists for Southampton, Leicester and Winchester, it would seem that their undefended state was as much to do with the attitude of the town authorities to the poorest members of their society as it was with any improvement in the security of the countryside.

The survival of town defences is variable and in many cases it is difficult to track down long stretches of wall. Turrets and gates have a better survival record, often because of their sheer size or their suitability as town gaols down to the nineteenth century, but many have been destroyed within the past century as increasing traffic demands have seen them as a constriction to traffic. At Canterbury, Lincoln and Southampton massive gateways survive, either isolated on a traffic island or still at peril to passing traffic; Canterbury's buses had to be specially designed to negotiate the narrow gateways of West Gate. Of the long defence line of Norwich (see Figure 6.6), which together with the river Wensum stretched for over two-and-a-half miles, there remain only short pieces of wall and isolated towers (followed in part by an inner ring-road scheme), and roundabouts and traffic lights at junctions are all that mark the site of the town's former gateways. London's wall has suffered badly through the centuries, buried in the dense concentration of buildings that made use of it as a convenient back wall, but the clearing of bombed sites and more recent redevelopment within the City has exposed some stretches, of which the Barbican site is now the best example. Very few towns have lost all trace of their walls, although those of Barnstaple and Doncaster have nothing to show on the ground. Field-work may often help in the identification of a lost ditch and some defences are still remembered in street names. Towns are usually well-mapped features, beginning with Speed's early seventeenth-century plans in many instances, and most can be explored with the aid of a copy of a fine eighteenth-century map. What is most important to remember is that long after their defensive function had ceased the town's defences continued in use as parish and other administrative boundaries.

Like the villages, the towns suffered a set-back in the second half of the medieval period. There is little evidence for the continuing expansion of suburbs during the fifteenth century. Those towns that relied upon wool manufacture, and that involved a great number, were seriously affected by the recession that bought about a decline in the export trade and the growing competition of a rural-based industry.[76] The plagues that had been inflicted upon the whole country in the fourteenth century became a phenomenon of the towns in the fifteenth, and this sapped them of their energy and their population. The Tudor topographer, Leland, described Norwich as a city in decay. Late medieval Lincoln had twelve of its 46 churches in a state of disrepair.[77] Everywhere buildings were neglected, streets left unpaved and the townspeople tried to evade the onerous taxations that were levied to cover the effects of disastrous fires which became an increasing threat to the late medieval urban population, cramped together in their timber-framed houses.

The decline of many medieval markets by 1500 is partly symptomatic of this general decline. Staffordshire's 45 markets had fallen to 20. Not until the Great Rebuilding of the late sixteenth-early seventeenth century did many towns revive, although some would inevitably never regain their medieval glory and would sink back into rural backwaters. Few were to go back to mere village status, but of those where the woollen industry had boosted them to small towns and then abandoned them as quickly as it had arrived the pattern of change was the most dramatic. Lavenham in Suffolk is probably the most complete late medieval townscape to remain in England, its parish church, guild halls and market-place all signs of its former status. It reflects better than most the fine balance to be sought in the landscape between medieval village and town.

Notes

1. See, for example, M.W. Beresford, *New Towns of the Middle Ages* (Lutterworth Press, Guildford, 1967); and M.W. Beresford and J.G. Hurst, *Deserted Medieval Villages* (Lutterworth Press, Guildford, 1971).

2. See M. Aston and J. Bond, *The Landscape of Towns* (Dent, London, 1976); and T. Rowley, *Villages in the Landscape* (Dent, London, 1978).

3. Beresford, *New Towns of the Middle Ages*, pp. 273–81.

4. The most satisfactory attempt is made by M.W. Beresford and H.P.R. Finberg, *English Medieval Boroughs: a hand-list* (David and Charles, Newton Abbot, 1973).

5. R.A. Dodgshon and R.A. Butlin (eds.), *An Historical Geography of England and Wales* (Academic Press, London, 1978), p. 106.

6. Ibid., pp. 142, 192.

7. W.G. Hoskins, *The Making of the English Landscape* (Hodder and Stoughton, London, 1955), p. 224.

8. H.C. Darby and I.S. Maxwell (eds.), *The Domesday Geography of Northern England* (Cambridge University Press, Cambridge, 1962), pp. 59-71, 139-50, 212-21.

9. R. Millward and A. Robinson, *The Lake District* (Eyre-Methuen, London, 1970), pp. 151-2.

10. Between 30 and 40 settlements were placed wholly or partly under forest law. See H.C. Darby and E.M.J. Campbell (eds.), *The Domesday Geography of South-East England* (Cambridge University Press, Cambridge, 1962), pp. 324-38.

11. H.C. Darby, *Domesday England* (Cambridge University Press, Cambridge, 1977); Darby *et al.*, *The Domesday Geography of England* (Cambridge University Press, Cambridge); *Eastern England* (1952); *Midland England* (1954); *Northern England* (1962); *South-East England* (1962); *South-West England* (1967).

12. S.E. West, 'The Anglo-Saxon Village of West Stow', *Medieval Archaeology*, vol. 13 (1969), pp. 1-20.

13. P.V. Addyman *et al.*, 'Anglo-Saxon Houses at Chalton, Hampshire', *Medieval Archaeology*, vol. 16 (1972), pp. 13-31; P.V. Addyman and D. Leigh, 'The Anglo-Saxon Village at Chalton, Hampshire: second interim report', *Medieval Archaeology*, vol. 17, (1973), pp. 1-25.

14. M. Bell, 'Excavations at Bishopstone', *Sussex Archaeological Collections*, vol. 115 (1977).

15. This important site remains to be fully published. For progress see *Medieval Archaeology*, vol. 21 (1977), p. 206.

16. *Medieval Archaeology*, vol. 21 (1977), pp. 212-13.

17. Royal Commission on Historical Monuments, *An Inventory of the Historical Monuments in the County of Northampton, Vol. II, Archaeological Sites in Central Northamptonshire*, (HMSO, London, 1979), pp. xlvii-liii.

18. See for example G.R.J. Jones, 'Settlement Patterns in Anglo-Saxon England', *Antiquity*, vol. 35 (1961), pp. 221-32; 'Basic Patterns of Settlement Distribution in Northern England', *Advancement of Science*, vol. 18 (1961), pp. 192-200; 'The Multiple Estate as a Model Framework for Tracing the Early Stages in the Evolution of Settlement' in F. Dussart (ed.), *L'Habitat et les Paysages Ruraux d'Europe* (Liège, 1971), pp. 251-67.

19. P.G. Bigmore, 'Suffolk Settlement — a Study in Continuity', unpublished PhD thesis, University of Leicester, 1973, p. 187.

20. M.W. Beresford and J.G. Hurst, 'Wharram Percy: a Case Study in Microtopography' in P.H. Sawyer (ed.), *English Medieval Settlement* (Edward Arnold, London, 1979), pp. 52-85.

21. C.C. Taylor, 'Polyfocal Settlement and the English Village', *Medieval Archaeology*, vol. 21 (1977), pp. 189-93.

22. Royal Commission on Historical Monuments, *An Inventory of the Historical Monuments in the County of Northampton, Vol. II*, pp. 37-9.

23. P. Wade-Martins, 'The Development of the Landscape and Human Settlement in West Norfolk from 350-1650 AD with particular reference to the Launditch Hundred', unpublished PhD thesis, University of Leicester, 1971. See also Wade-Martins, 'The Origins of Rural Settlement in East Anglia' in P.J. Fowler (ed.), *Recent Work in Rural Archaeology* (Moonraker Press, Bradford-on-Avon, 1975), and 'Village Sites in Launditch Hundred', *East Anglian Archaeology*, Report No. 10 (1980).

24. Faden's Map of Norfolk, 1797, reprinted as *Norfolk Record Society*, vol. XLII (1975).

25. Wade-Martins, 'The Origins of Rural Settlement in East Anglia'.

26. H. Thorpe, 'The Green Villages of County Durham', *Trans. Inst. Brit. Geog.*, vol. 15 (1949), pp. 155–80.

27. B.K. Roberts, 'Village Plans in Co. Durham: A Preliminary Statement', *Medieval Archaeology*, vol. 16 (1972), pp. 33–56; 'Rural Settlement in County Durham: Forms, Pattern and System' in D. Green, C. Haselgrove and M. Spriggs (eds.), *Social Organisation and Settlement*, Part II, BAR International Series (supplementary), vol. 47, no. ii (1978), pp. 291–322.

28. See, for example, P. Allerstone, 'English Village Development: Findings from the Pickering district of N. Yorkshire', *Trans. Inst. Brit. Geog.* vol. 51 (1970), pp. 95–109: J.A. Sheppard, 'Pre-inclosure Field and Settlement Patterns on an English Township – Wheldrake, near York', *Geografiska Annaler*, vol. 48B (1966); 'Metrological Analysis of Regular Village Plans in Yorkshire', *Agri. Hist. Rev.*, vol. 22 (1974), pp. 118–35; 'Medieval Village Planning in northern England: some evidence from Yorkshire', *Journal of Historical Geography*, vol. 2, no. 1 (1976), pp. 3–20.

29. Roberts, 'Village Plans in Co. Durham', p. 53.

30. Sheppard, 'Medieval Village Planning in Northern England'.

31. Sheppard, 'Pre-inclosure Field and Settlement Patterns in an English Township – Wheldrake, near York'.

32. S. Goransson, 'Regular Open-field Pattern in England and Scandinavia solskifte', *Geografiska Annaler*, vol. 43 (1961), pp. 80–104.

33. Allerston, 'English Village Development', pp. 102–3.

34. See for example J. Thirsk, 'The Common Fields', *Past and Present*, no. 29 (1964), pp. 3–25.

35. C. Taylor, 'The Making of the English Landscape: 25 years on', *The Local Historian*, vol. 14 (1980), pp. 195–201.

36. The debate is well illustrated in M.M. Postan, 'Some Economic Evidence of the Declining Population of the later Middle Ages', *Economic History Review*, vol. II (1949–50), pp. 221–46; *The Medieval Economy and Society* (Weidenfeld and Nicolson, London, 1972); S.M. Wright, 'Barton Blount: Climatic or Economic Change; an addendum', *Moated Site Research Group*, Report No. 7 (1980), pp. 43–6.

37. G. Beresford, 'Three Deserted Medieval Settlements on Dartmoor: A Report on the late E. Marie Minter's Excavations', *Medieval Archaeology*, vol. 23 (1979), pp. 93–158.

38. See G. Beresford, *The Medieval Clay-Land Village: Excavations at Goltho and Barton Blount* (Society for Medieval Archaeology, Monograph Series No. 6, London, 1975).

39. A.R.H. Baker, 'Evidence in the "Nonarum Inquisitiones" of Contracting Arable Lands in England during the early Fourteenth Century', *Economic History Review*, vol. 19 (1966), pp. 518–32.

40. A.R.H. Baker, 'Contracting Arable Lands in 1341', *Bedfordshire Historical Record Society*, vol. 49 (1970), pp. 7–18.

41. K.J. Allison, 'The Lost Villages in Norfolk', *Norfolk Archaeology*, vol. 31 (1955), pp. 116–62.

42. M.W. Beresford, *The Lost Villages of England* (Lutterworth Press, Guildford, 1954); and Beresford and Hurst, *Deserted Medieval Villages*.

43. H. Thorpe, 'The Lord and the Landscape', *Transactions of the Birmingham Archaeological Society*, vol. 80 (1965); reprinted in D.R. Mills (ed.), *English Rural Communities* (Macmillan, London, 1973), pp. 31–82.

44. The village is well described in W.G. Hoskins, *English Landscapes*, (BBC Publications, London, 1973), pp. 46–7.

45. Beresford and Hurst, *Deserted Medieval Villages*, pp. 104–5.

46. P.G. Bigmore, *The Bedfordshire and Huntingdonshire Landscape* (Hodder and Stoughton, London, 1979), pp. 84, 127.

47. Beresford and Hurst, 'Wharram Percy: a Case Study in Microtopography'.

48. Beresford, *The Medieval Clay-Land Village: Excavations at Goltho and Barton Blount*.

49. P.A. Rahtz, 'Holworth, Medieval Village Excavation 1958', *Proc. Dorset Nat. History and Archaeological Society*, vol. 81 (1959), pp. 127–47.

50. Aston and Bond, *The Landscape of Towns*.

51. Beresford, *New Towns of the Middle Ages*.

52. M. Biddle and D. Hill, 'Late Saxon Planned Towns', *Antiquaries Journal*, vol. 51 (1971), pp. 70–85. Also by Biddle, 'The Evolution of Towns: Planned Towns before 1066' in M.W. Barley (ed.), *The Plans and Topography of Medieval Towns in England and Wales* (CBA Research Report, No. 14, 1976), pp. 19–32.

53. Bigmore, *The Bedfordshire and Huntingdonshire Landscape*, p. 98.

54. Darby *et al.*, *The Domesday Geography of Eastern England*, p. 198.

55. Bigmore, *The Bedfordshire and Huntingdonshire Landscape*, pp. 106–8.

56. L.M. Munby, *The Hertfordshire Landscape* (Hodder and Stoughton, London, 1979), pp. 100–1.

57. Beresford, *New Towns of the Middle Ages*, pp. 457–9.

58. Ibid., pp. 426–8. There has been some confusion over the location as two 'Newton' names appear on the modern Ordnance Survey maps. Recent investigation suggests the western Newton as the more likely site, as it lies at the southern end of Newton Bay where deep water allowed vessels to come in.

59. Ibid., pp. 513–14.

60. But see N. Scarfe, *The Suffolk Landscape* (Hodder and Stoughton, London, 1972), pp. 205–8.

61. J.R. Smith, *Towns of Essex, Chelmsford* (Essex Record Office Publications, No. 57, n.d.). Nineteen markets were listed for Essex and, although the list appears to have excluded some markets known from other sources, the overall evidence is for late medieval decline.

62. Aston and Bond, *The Landscape of Towns*, p. 98.

63. M.R. Conzen, 'Alnwick, Northumberland: A Study in Town Plan Analysis', *Institute of British Geographers*, vol. 27 (1960), pp. 3–122.

64. For further details of Glastonbury, see M. Aston and R. Leech, *Historic Towns in Somerset* (Committee for Rescue Archaeology in Avon, Gloucestershire and Somerset, Surveys No. 2, Bristol, 1977).

65. *Current Archaeology*, vol. 68 (1979), pp. 280–4.

66. The situation is admirably summarised in D.M. Palliser, 'Sources for Urban Topography: Documents, Buildings and Archaeology' in M.W. Barley (ed.), *The Plans and Topography of Medieval Towns in England and Wales*, pp. 1–7.

67. Royal Commission on Historical Monuments, volumes for York, 4 vols., (HMSO, London, 1972, 1973, 1975, 1977); Stamford (HMSO, London, 1977); and Salisbury (HMSO, London, 1980).

68. C. Platt, *Medieval Southampton* (Routledge, London, 1973). In the twelfth century the town was almost exclusively timber-framed, but stone became more common, as elsewhere, partly as a precaution against fire.

69. *Current Archaeology*, no. 56 (1977), pp. 280–1.

70. The intrusion of eighteenth- and nineteenth-century cellars into medieval layers in the centre of small towns has virtually destroyed anything of archaeological significance. See, for example, Aston and Leech, *Historic Towns in Berkshire: An Archaeological Appraisal* (Berkshire Archaeological Committee, Reading, 1978).

71. H. Forrester, *Timber-framed Building in Hertford and Ware* (Hertfordshire Local History Council, Hitchin, 1965).

72. Hoskins, *The Making of the English Landscape*, pp. 227–8.

73. M.W. Barley, 'Town Defences in England and Wales after 1066' in M.W. Barley (ed.), *The Plans and Topography of Medieval Towns in England and Wales*, pp. 57–71.

74. Platt, *Medieval Southampton*, p. 39.

75. D.J. Keene, 'Suburban Growth' in M.W. Barley (ed.), *The Plans and Topography of Medieval Towns in England and Wales*, pp. 71–82.

76. C. Platt, *The English Medieval Town* (Secker and Warburg, London, 1976), pp. 85–95.

77. A.R.H. Baker, 'Changes in the Later Middle Ages' in H.C. Darby (ed.), *A New Historical Geography of England before 1600* (Cambridge University Press, Cambridge, 1976), pp. 186–247; and Sir Frances Hill, *Medieval Lincoln* (Cambridge University Press, Cambridge, 1948), pp. 286–7.

7 ROADS AND TRACKS

Brian Paul Hindle

Introduction

It is curious that whilst so much has been written by economic historians and historical geographers about the medieval economy and about trade in particular, very little had been written about the means by which that trade was carried on. What little has been written about roads is generally confined to maintenance, the means and safety of travel, and to the state of the roads. Few have attempted to view the roads as a system. The reason for this neglect is not difficult to ascertain, for it is almost certainly related to the paucity of the evidence available. In archaeological terms it is difficult to date a road, or to say when it was or was not in use; in terms of written evidence there is little indication either in the form of maps, itineraries or written descriptions to suggest where the roads were. What is certain, however, is that the road system must have formed a vital part of the medieval economy, aided in certain places by river transport, especially for the movement of bulky goods. And yet virtually no new roads were built between the Roman and Turnpike eras; here lies the core of the problem — where were the roads and how did they come into existence? This chapter must thus devote a substantial amount of space to the sources of information.

Previous research can be summed up under six headings. First, the demand for roads has been clearly demonstrated even by those writers who confine their comments to the growth of the economy, and either do not mention roads at all, or give them a cursory single sentence.[1] Traffic included merchandise carried both on horseback and by cart as well as various officials, both secular and ecclesiastical, pilgrims, justices and tax collectors; medieval men often travelled beyond their village or town and mobility increased throughout the period.[2] In early medieval times the king and his court also moved incessantly around the kingdom, occasionally with the army. Second, it is clear that many of the engineered Roman roads remained intact and, with the growth in trade, were used increasingly.[3] Third, although there was virtually no road-building as such during medieval times, new roads grew from habitual lines of travel; these were the roads which in C.T. Flower's memorable phrase 'made and maintained themselves'.[4] There is thus a fundamental distinction between the surviving Roman roads and these later routes. Fourth, it is important to remember that the medieval road was regarded more as a right of way than as a physical track; the traveller had the right to diverge from the road if it was impassable, even to the extent of trampling crops.[5] Fifth, it is now becoming clear that the road system was inadequate for the amount of traffic, even in winter, and that news, in particular, could travel remarkably rapidly.[6] Last, it seems that where possible river transport was used mainly for the movement of bulky merchandise, with the network of roads acting as feeders to the river system.[7]

There have been only two previous attempts to make a comprehensive study of the road network of medieval England. The first was by F.M. Stenton in 1936, and part of it was reprinted with the facsimile of the Gough Map;[8] surprisingly, he was the first to use the evidence of the roads shown on this map. Stenton notes the tradition surviving into Norman times that the four 'great roads' (Watling Street, Ermine Street, Fosse Way and Icknield Way) were under the king's special protection, and that they had probably remained in use throughout the Dark Ages. Under Norman and Angevin rule the centralisation of government in London meant that the Roman road system once again became vital, as it too had been centred on London. He was also the first to point out the remarkable mobility of the itinerant royal court: 'in every part of England there existed a network of local roads which could bear the passage of a very considerable company, and the transport of provisions for its entertainment'.[9] His conclusion is worth quoting at some length:

> The road system lacked both the directness and the material definition of the Roman achievement . . . Nevertheless, with all its defects the road system of medieval England provided alternate routes between many pairs of distant towns, united port and inland market, permitted regular if not always easy communication between the villages of a shire and the county town . . . and brought every part of the country within a fortnight's ride of London. In the last resort it proved not inadequate to the requirements of an age of notable economic activity, and it made possible a centralisation of national government to which there was no parallel in western Europe.[10]

The most recent study of the road system is by the present author, and it attempts a cartographic synthesis of the main types of evidence available in order to produce a map of known medieval roads; a similar approach will be used and extended later in this chapter.[11] One recent book deserves a mention here; it is the valuable set of aerial photographs by M.W. Beresford and J.K.S. St Joseph, which has a short section on roads, but which also deals with local roads in town and villages at many points throughout the book.[12]

Archaeological and Documentary Evidence

Because most medieval roads were not formally constructed or engineered, it is virtually impossible to say when any particular road was or was not in use; many tracks across fields have been ploughed up, often after enclosure, whilst others have had modern roads built over them. This contrasts with the reconstruction of the Roman road system where the main evidence has always been archaeological. A few new roads were built — in 1277 and 1283, for example, roads and passes into Wales were enlarged and widened for Edward I's campaigns — and perhaps most notable were the causeways built across the Fenlands. The building and upkeep of bridges, however, attracted much more attention and records are much more detailed; unfortunately they tell us little of the roads leading to them.

One of the standard texts on medieval archaeology makes only a few passing references to roads, chiefly in connection with deserted village sites where the whole medieval landscape has been fossilised.[13] Field evidence leaves us nothing but a host of short sketches of disjointed local tracks and holloways (sunken roads) which whilst being of some importance in studies of individual settlements, gives us little idea which way medieval man travelled from village to village or from town to town. Equally parochial are studies of roads in the towns themselves.[14]

Documentary evidence is equally sparse and tends to emphasise only the difficulties of medieval road travel, as evidenced by C.T. Flower's collection of medieval court cases concerning public works.[15] Bridges figure more prominently than roads, though there are cases concerning drainage ditches and various blockages of the highway. A notable example was that in 1386 the Abbot of Chertsey allowed two wells (*sic*), 12 feet wide and 8 feet deep, to exist in the high road from Egham to Staines; a man had drowned in one of them and the Abbot had claimed his goods.[16] The fact that the *Canterbury Tales*, which is woven around a pilgrimage from Southwark to Canterbury, makes no reference either to the road or to the state of the road might be taken to imply that the route was well known and in good condition.

Cartographic Evidence

The paucity of archaeological and written evidence is happily not matched by that from contemporary maps. Matthew Paris (*fl.* 1250) was a monk at St Albans and drew four maps of Britain which are based on an itinerary from Dover to Newcastle, a route which forms the backbone of his maps.[17] The route goes by way of Canterbury, Rochester, London, St Albans, Dunstable, Northampton, Leicester, Belvoir (a cell of St Albans), Newark, Blyth, Doncaster, Pontefract, Boroughbridge, Northallerton and Durham; J.B. Mitchell suggests that it continues to Berwick, at least on the third map (Figure 7.1).[18] Matthew Paris probably derived this route from a written itinerary, and the maps represent an original attempt to depict oral and written information in cartographic terms. However, the maps are crude and must be used with caution, for a legend on the fourth map disarmingly states that the island would have been elongated if the page size had been larger.

Much better evidence for the road system, however, is to be found on the Gough Map of about 1360 which depicts some 2,940 miles of roads covering most of England.[19] Interpretation of this map is made difficult because neither its purpose nor its sources are known, but it appears to have been an official compilation for government use, perhaps amended for use in certain areas; for example the extant copy has networks of local roads in south-east Yorkshire and Lincolnshire.[20] Almost 40 per cent of the routes shown are along the line of Roman roads (see Figure 7.1), and they have been described and classified elsewhere.[21]

The map omits several well-known roads such as those from London to Dover and from York to Newcastle, although it does show the towns *en route* correctly. The omission of the latter route is a particular puzzle, as it was much used as the main route to Scotland. Viewing the map as a whole, the sheer number of towns shown would enable the traveller to plan a journey, even if an actual route was not shown. Many places are included only because they were stages between larger towns; Bitchfield (between Stamford and Lincoln) and Bentham (between Settle and Lancaster) are such places, although the criteria for the choice of towns shown on the map remain a total mystery. The roads do, however, reflect the centralisation of government, for there is

Figure 7.1: The roads of Matthew Paris and the Gough Map

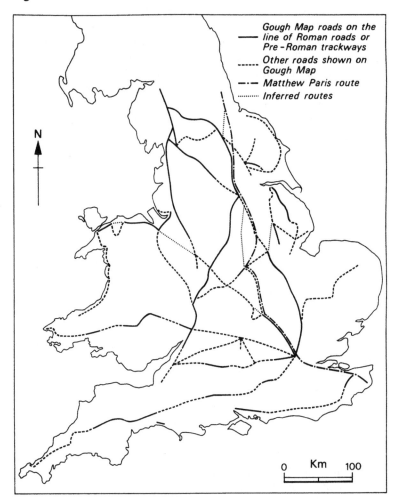

clearly a national road system radiating from London, although certain
important towns, such as Plymouth, King's Lynn and Colchester which
were probably among the ten largest in the country, are not connected
to the network at all. York, the second largest town, is poorly served,
and Lincoln has only local roads, although here the normal route to
York, for example, would have been by river.

It is tempting to presume that the lines marked on the map actually
represented roads on the ground; as far as those routes along Roman

roads are concerned, this would seem to be a fair assumption, although there are numerous examples of where later roads have developed alongside Roman roads, because travellers shunned the hard surface (if it still survived) for the softer ground alongside. As for the routes which do not follow Roman roads, these must have been tracks which made themselves through the continual passage of traffic; at worst they were directions on the maps to guide the traveller across open country.

The direct historical evidence for medieval roads is thus very poor, and reliance has to put on other more indirect sources, such as itineraries, in order to complete the picture.

Itineraries

The various itineraries of the medieval period, whether compiled at that time or posthumously, provide direct evidence of the movement of individuals. There is, for example, the itinerary of Giraldus Cambrensis who toured Wales in 1188 with Archbishop Baldwin,[22] and a whole host of bishops' itineraries from the mid-thirteenth century onwards.[23] These ecclesiastical itineraries are of limited use, as the bishops visited a very biased selection of places, and in most cases the itineraries are both parochial and lacking in detail. Records also exist of the movements of other individuals, such as those of the warden and fellows of Merton College, Oxford,[24] and of Robert of Nottingham who was buying wheat for the king in 1324-5 in the area around the river Trent; he appears to have had an early edition of the Gough Map with him, as he gives the same mileages as those on the map.[25]

The most complete itineraries available are those which have been compiled for the kings of England; the kings were itinerant, almost constantly on the move, and could be expected to visit a much less biased selection of stopping places than anyone else, whether castles, manors, religious houses or market towns. Moreover, as the entire court moved with the king, adequate roads were clearly necessary; even in the time of King John 'the essentials of government, the *hospitium regis* . . . followed the court; a train of from ten to twenty carts and wagons'.[26]

Once the places named in the itineraries have been located accurately, two problems remain. The first is that there are gaps in the itineraries which we are unable to fill. For example, as the first of the following extracts shows, we have no idea of King John's whereabouts for ten days, during which time he traversed part of England about whose roads little is known; indeed he may have gone by boat, but in that case he would hardly have taken so long. The short extract from Edward I's itinerary presents another intractable problem, for he may have journeyed out from Lyndhurst on 11th December 1289. The second problem is that the king may have travelled by river, eschewing the roads entirely, and this certainly happened on occasion on the Thames and on the Trent and Ouse, particularly between Lincoln and York.

Itinerary of 7[a] *John*[27]			*Itinerary of 18*[b] *Edward I*[28]		
1206	February 20	Carlisle	1289	December 10	Lyndhurst
				11	—
	March 1	Chester		12	Lyndhurst

Notes:
a. 7th year of John's reign: May 1205–May 1206.
b. 18th year of Edward's reign: November 1289–November 1290.

Despite these problems, it is reasonable to suppose that if one or more of the kings for whom we have day-to-day details of their movements used particular lines of travel frequently, then some kind of road or track is certain to have existed. A study has been made of the itineraries of John, Edward I and Edward II (Figures 7.2a and b) covering the period from 1189 to 1327, and a more detailed description is given elsewhere.[29] When the routes taken by the three monarchs are compared, a concentration emerges in the area between London, Northampton, Worcester and Dorchester, with two linear routes, one north to York and Berwick, the other to Canterbury and Dover. Figure 7.3 includes all routes travelled four or more times by more than one of the three kings. It can be seen that certain areas of the country were rarely visited by monarchs; Wales, the north-west, the south-west and East Anglia stand out in particular, and it is particularly difficult to explain why the monarchs avoided the last two of these areas, which had such great economic importance.

Commercial Evidence

Whereas the itineraries relate to the movement of individuals, even if they travelled with the whole paraphernalia of the royal court, different criteria must apply to the movement of bulky produce. Little detailed

Figure 7.2a and b: Itinerary of Edward I

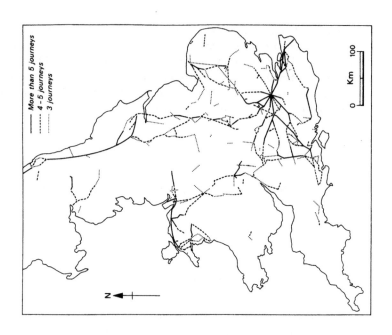

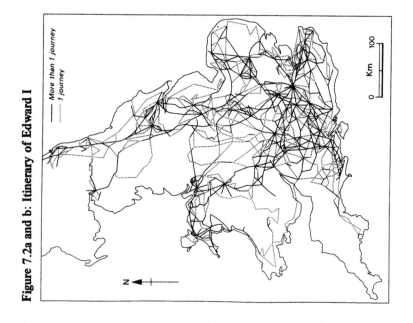

Figure 7.3: Minimum aggregate road network based on the royal itineraries

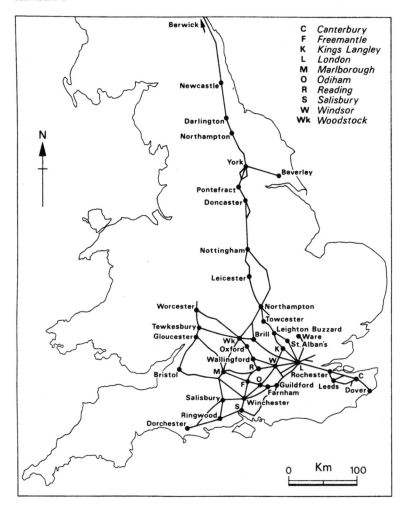

evidence of the movement of produce on any large scale survives from the Middle Ages, and it was only when the government became involved in moving produce that records were kept and have survived. When the king had to provision his armies for campaigns in Wales, Scotland or France, he sent out officials to purchase or seize whatever was required, and to arrange for its transportation. The accounts which were kept give details, amongst other things, of the type of transport (that is, whether by river or road), and can be interpreted to reveal the routes taken.[30] Once again, the evidence does not give national coverage, but it is particularly good for Lincolnshire and Yorkshire. The general pattern seems to have been that goods were moved by road from smaller to larger centres, and eventually by river to the customs ports such as Boston or Hull. In the latter case, 'whilst roads definitely were available, transport to the port was almost exclusively by river'.[31] Obviously these two counties are well served by rivers, and such a reliance on rivers could not apply to most of England, where individual merchants would have moved from town to town with their goods carried by pack-horse or cart. There are numerous other references to the movement of produce; R.A. Donkin notes the movement of wool from various Cistercian houses to distant ports, for example from Holm Cultram (Cumberland) to Newcastle, Furness (Lancashire) to Beverley, Basingwerk (Flint) and Vale Royal (Cheshire) to London, and Combermere (Cheshire) to Boston. It seems that most of the wool was carried by cart, and these long hauls imply that the roads were not unduly difficult; yet we know virtually nothing of the routes taken.[32] Similarly, from the detailed records which survive for the village of Cuxham (Oxfordshire), we know that hay, stone and wood were supplied to the demesne from numerous places, some over ten miles distant; however, we know nothing of their transportation, nor of how the produce of the village was sent to Abingdon and Henley.[33]

In order to attempt a reconstruction of the amount of commercial traffic between the towns of England, a theoretical approach has been used, in the form of a gravity model which, following Newtonian physics, suggests that the amount of interaction between every pair of towns is proportional to the product of their population and inversely proportional to the square of the distance between them. Thus by multiplying together the population of each pair of towns and dividing the result by the square of the distance between them, we can obtain some idea of the relative importance of the roads of England, and in particular we can see where roads ought to have existed to cater for commercial traffic, but which are not evident from the strictly acceptable historical evidence. There are two main problems with such an approach, the more important being the difficulty of obtaining accurate borough population figures and of deciding how many boroughs to include.[34] Less difficult is the problem of plotting the interactions on a map. The process is described elsewhere, and the result for the year 1348 is shown in Figure 7.4.[35] The routes, which

Figure 7.4: Theoretical routes network, 1348

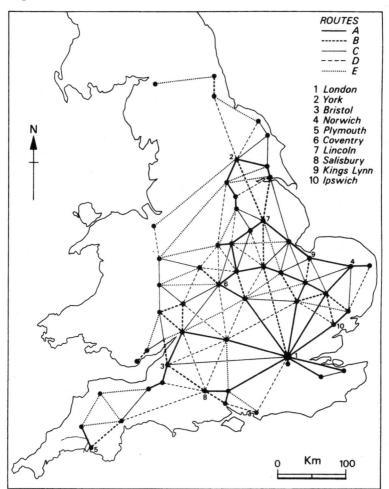

ROUTES
— A
------- B
— C
---- D
.... E

1 London
2 York
3 Bristol
4 Norwich
5 Plymouth
6 Coventry
7 Lincoln
8 Salisbury
9 Kings Lynn
10 Ipswich

N

0   Km   100

are roughly divided into five groups, present the appearance of an ordered rather than a random network. There are gaps along the south coast where the ports do not figure in the population totals, and the well-used routes from London to Windsor and Reading are missing for the same reason.

Such an approach is based firmly on the notion of an integrated economy with a large amount of commercial interdependence between the towns of various sizes which served the agricultural and urban systems. The larger the town, the greater was its amount of trade and the greater its sphere of influence.

The Road Network

From the various types of evidence already presented it is now possible to build up a picture of the road network that existed in medieval times. Our starting-point must be the Roman road system, large parts of which were clearly still in use, as shown by the Gough Map and the royal itineraries, despite the ravages of time and weather since the departure of the Romans. Indeed their routes have persisted, and many are still in use today, though not always as through routes. All this evidence has been brought together to show which Roman roads were still in use in medieval times.[36] To this must then be added the evidence of roads on contemporary maps and the major routes used by the kings which were not along Roman routes, all of which can be truly described as the medieval roads which made and maintained themselves, as opposed to the Roman roads still in use (Figure 7.5).

This distinction between the two types of road is especially apparent in certain areas: Cirencester is an important junction in the Roman system, yet Oxford and Windsor are totally divorced from it. Other distinctions emerge too: Oxford is an important junction on the map evidence, but the itineraries give more prominence to the palace at Woodstock, eight miles to the north. The major centre is clearly London, followed, in order, by York, Marlborough, Leicester, Salisbury, Winchester, Woodstock, Lincoln, Chester, Shrewsbury, Lichfield, Gloucester, Oxford and Windsor. This network is clearly at its best in central and southern England, and poor in all the surrounding areas. To determine more accurately where the most significant gaps occur we can add the theoretical route network (Figure 7.4) to that of the known medieval roads (Figure 7.5), and the result appears as Figure 7.6. Where the theoretical interactions fit the known network, they are shown as solid lines, divided into three classes in order of importance. Where there are no known roads to fit the interactions, then dashed lines appear, and it is clear that strict historical evidence for roads is lacking in several areas, notably in East Anglia, the south-west and towards the Welsh borders. In particular, large towns such as Plymouth, King's Lynn, Hull and Yarmouth are not connected at all to the known road network, and it is to such

Figure 7.5: Known medieval roads

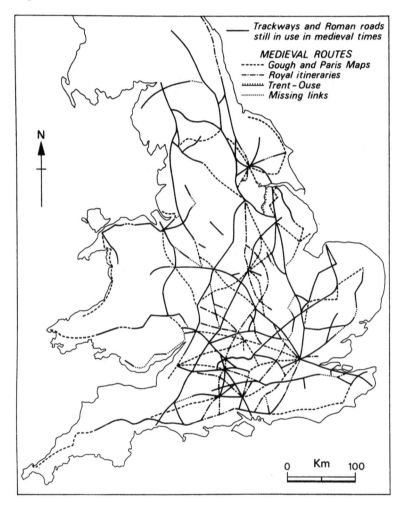

areas and towns that more local research must be devoted to determine what road links existed in medieval times.

Local Studies

Local studies fall into one of three categories: they may deal with the roads in a particular parish, they may look at an individual road between

Figure 7.6: Theoretical routes on the known medieval road network

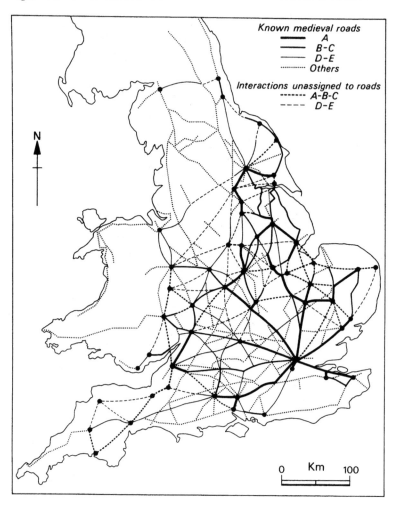

two places, or they may attempt a synthesis of the road network in a larger areas such as a county.

As an example of the first, we can turn to the final section of Beresford and St Joseph's aerial survey of medieval England.[37] Their introductory remarks stress the importance of local roads: 'Journeys to markets, churches and coasts are the principal exceptions to the generalisation that most medieval roads were entirely local in purpose with an ambition no higher than to serve the villagers' immediate wants'.[38] Most such roads would not continue to the adjacent parishes, but instead would lead out into the fields or to the outlying woods, pasture or mills. Such tracks would have been extended out from the villages as required, and had a very irregular appearance in both their course and width. Such patterns appear only in areas of nucleated settlement and open fields; in areas of dispersed settlement in the western counties this radical network is replaced by one with lanes linking one isolated farm to another, though with still few direct routes from one village to the next. Beresford and St Joseph are much concerned with the changes that took place during enclosure, noting in particular the relative freedom of medieval roads to move or spread out, compared to enclosure roads which were confined by hedges. However, the enclosure surveyor did sometimes keep the furlong boundaries, and the right-angled bends in roads across enclosed land often still reflect the medieval ploughing pattern.

Examples of the way in which a medieval road would spread out over a large area are best seen where the road left cultivated land, and in particular where it had to climb a hill. On Twyford Down, near Winchester, the road becomes a fan of numerous tracks ascending the 200 feet climb of the hill; another good example can be seen at Postern Hill, south of Marlborough.[39] There are numerous other examples throughout the country, but they are limited in areal extent, and in any case it is impossible to be dogmatic about the date of such tracks.

Beresford and St Joseph choose the village of Padbury (Buckinghamshire) to study in some detail, having already looked at its fields and the effects of enclosure. Their starting-point is an Elizabethan plan of the parish dated 1591 (Figure 7.7).[40] This shows a radial network of lanes with some linking cross-routes, leading out into the fields from the village street. Most of them degenerate into smaller tracks, including the one named 'Buckingham Waye' which leads to the county town, less than three miles distant. The other route named as heading for a specific place, 'Whadden Waye', made its way across the fields for almost six miles to Whaddon, but this route is now only a footpath.

Roads played an important role in determining the success or otherwise of towns and their markets; many of the towns which failed were simply not well located with respect to the road system. At Brough in Westmorland, for example, a castle was built in about 1100 on the site of an old Roman fort with a church on an adjacent mound. However, the foundation failed to prosper, almost certainly because markets were held further north at Market Brough, which was sited

Figure 7.7: Padbury Parish, Buckinghamshire, in 1591

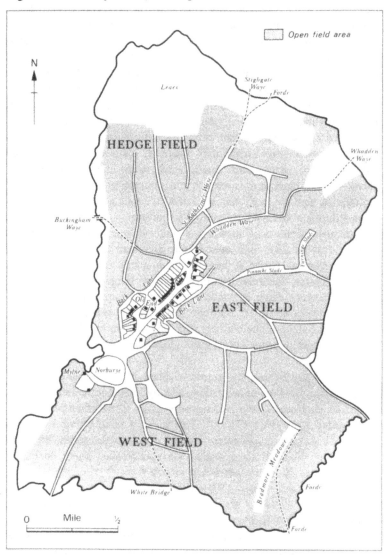

Source: Based on M.W. Beresford and J.K.S. St Joseph, *Medieval England: An Aerial Survey* (Cambridge University Press, Cambridge, 1979).

on a slight medieval diversion from the Roman road descending from Stainmore.

In larger towns, too, the presence of roads often had a marked effect on the town plan. In Ludlow (Salop), for example, the town was originally laid out between the castle, which is on a rocky promontory above the river, and the main north–south route through the Marches; this junction (the Bull Ring) formed one end of a large market area, now much infilled. The town grew around the market area and on either side of the north–south route (Corve St–Old St), until a large extension to the town was laid out in the thirteenth century and traffic was diverted through the main market and down Broad Street to the new bridge over the river Teme. The road to Hereford, south of the river, may also have been diverted some distance to the west, although its earlier course is uncertain.

Studies of individual routes are few and far between; a recent example is a study of the road from Stamford (Lincolnshire) to Kettering (Northamptonshire) (Figure 7.8a).[41] In this case various subsequent diversions have been noted, both in the towns and villages *en route* and in the countryside. In particular there is a fine holloway some 6 feet deep and 35 feet wide overall south of Bulwick, which runs parallel to the modern road for most of the way to Deenethorpe where the medieval road runs into the village, and is represented by a holloway once again south of the cul-de-sac in the village.

A short distance to the south-east in Cambridgeshire is a somewhat more complex example showing the changes in a route from Roman times to the present day, where the Roman road was probably abandoned in favour of drier routes further west, most of which have, in their turn, been abandoned at least as through routes (Figure 7.8b).[42] The main medieval road from Wansford went by way of Coppingford and Ogerston to Alconbury. Ogerston was a medieval manor, and although a farm is all that remains today, it is marked on the Gough Map as the stage between Huntingdon and Wansford. Another medieval route leaves Ermine Street near Sawtry, also trying to avoid the Fenlands; each of the villages through which this old route passes are laid out along it, even though it is now represented by a disjointed set of lanes and paths. In addition, the Roman road was probably in use again later in the medieval period.

A special type of road in the form of causeways was required in the Fenlands, and these were the only large-scale road-building projects of medieval times. The cathedral city of Ely in particular had to be linked to *terra firma* by three causeways built in the eleventh and twelfth centuries.[43]

Figure 7.8a: Medieval roads between Bulwick and Deenethorpe, Northamptonshire

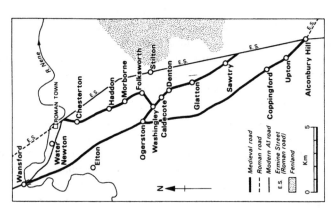

Figure 7.8b: Medieval roads between Alconbury Hill and Wansford, Cambridgeshire

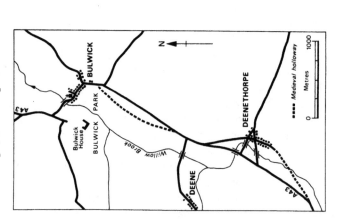

Source: Based on C. Taylor, *Roads and Tracks of Britain* (Dent, London, 1979).

Figure 7.9a: Routes derived from place-names in Cheshire

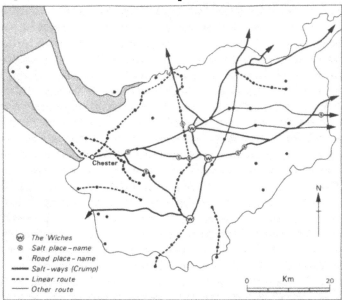

Figure 7.9b: Royal itinerary and Gough Map routes in Cheshire

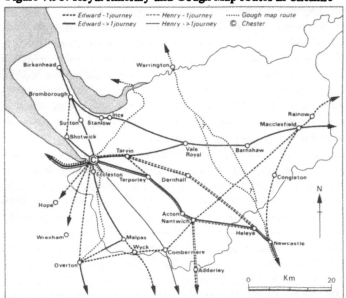

Source: Based on C. Taylor, *Roads and Tracks of Britain* (Dent, London, 1979).

As far as relatively large areas are concerned, there have been virtu-ally no attempts to look beyond particular roads or parishes to establish what the medieval road network was like. However, the present author has made a study of the Diocese of Carlisle which is fortunate in having the itinerary of a medieval bishop, in addition to all the map and royal itinerary evidence, to help in a reconstruction of the road system.[44] Here, however, it is proposed to look in detail at the county of Che-shire.[45] While this short study cannot hope to be definitive because of the sheer lack of field evidence, it attempts to show how the various sources and methods already described can be applied to such an area. As the only previous writing on medieval roads in Cheshire can hardly be described as comprehensive or systematic, we must start again from scratch.[46] The first step is to establish the network of Roman roads — though the precise route of large sections of it is uncertain — notably the route from Chester to Warrington. Settle-ments on this route contain place-name elements which support the existence of a road; they include 'ford', 'bridge', 'street' and 'stretton', and the Welsh elements 'cryw' (ford), 'fford or heol' (road) and 'sarn' (causeway).[47] These place-names appear in Figure 7.9a, where, by a process of connecting each one to its nearest neighbour and looking in detail at the alignments of the villages and their roads, it is possible to piece together several 'linear routes', one of which is along the Chester–Wilderspool route. D. Sylvester has also discovered three other routes in the east of the county which have been partly identified in the field. Cheshire was important commercially for its salt production and W.B. Crump has identified several 'salt-ways' linking the 'wiches' (Northwich, Middlewich and Nantwich) to various 'salt' place-names such as Saltersford.[48] Salt-ways were certainly not special roads in any sense, and Crump connects his far-flung 'salt' place-names by roads noted in W. Harrison and H.J. Hewitt.[49]

Place-names of course are only of limited value, principally because they do not give a date to a road, and are more likely to refer to pre-medieval roads in any case. They do, however, supplement our know-ledge of routes that may have been in use before medieval times. More useful are the royal itineraries (for example, the travels of Edward I and Henry III which are plotted in Figure 7.9b).[50] Many of the routes taken by the monarchs follow known roads, but some do not, for example those between Ince and Vale Royal and between Maccles-field and Nantwich. Edward's route to Overton probably went along the route shown on the Gough Map. Few of the routes used by the kings follow place-name roads; in fact only those from Nantwich to

Combermere and from Barnshaw to Macclesfield do so. The Gough-map roads are also shown on the maps, including the main road to the north which passes through the county from Newcastle to Warrington and although two intermediate places are shown its precise route is uncertain. The other road on the map was from Shrewsbury and divides to go into North Wales and to Liverpool; again, its route can only be guessed at.

Other roads must have served Chester, which was an important town in medieval England. Although it probably had a population of just over 2,000 for most of the period, its lack of growth meant that it slipped from being the fourth largest borough in England in 1086 to forty-second largest in 1348.[51] It would have had strong links to Shrewsbury, York and London and although the link to London is shown via Shrewsbury on Figure 7.4, it probably went by way of Newcastle under Lyme (Staffordshire).

In such local studies it is also sometimes useful to use post-medieval evidence, in order to confirm the existence of earlier roads. The earliest of such sources usually dates from the sixteenth and seventeenth centuries, and for Cheshire the roads given by Harrison, Smith, Ogilby and Morden date from this period.[52]

A synthesis of all this information, together with much detailed field-work, would lead us nearer a true picture of the medieval roads of Cheshire. As it stands, this cartographic approach has left us with many unresolved problems, in particular the routes taken by the Gough-map roads and the general confusion around Middlewich and Nantwich. In some areas, parallel lines of travel less than a mile apart follow each other over considerable distances, and there appears to be no major route into Wales from the southern half of the county. None the less, the maps show the importance of Chester and the main route from Newcastle under Lyme to Warrington. On the other hand, the 'wiches' are surprisingly poorly connected, although they are all linked to Chester, which may have been an important distribution centre. A generally flat county such as Cheshire would present few obstacles to travel, which may well account for much of the proliferation of routes.

Conclusion

This short study of Cheshire has attempted to show how all the sources and methods mentioned earlier in the chapter can be applied to an area the size of a county, and we must enter a plea for more research at this level. It is possible to deal with individual roads at this scale, and then to build up a broader picture than can be obtained by looking solely at one road. The method is relatively simple; the first step is to amass all the information possible, whether documentary, cartographic, linguistic or theoretical, and then to go out into the field to see what remains (if anything), and to see more directly what options were open to the traveller in medieval England.

In this chapter our main concern has been with the location of roads, but clearly there is another dimension — that of time. If the road network at its greatest extent is difficult to trace, then any attempt to see how it developed during the medieval period, or to try to map it for another particular year, is virtually impossible with the evidence available. Clearly, some Roman roads survived intact at the start of medieval times, and more roads, whether Roman or those which made themselves, came into use as the economy grew. The creation of new towns and the growth of certain industries would also have affected the pattern. There must have been no further growth of the system for many years after the Black Death (1348), as the drop in population would have reduced the need for roads. But how long this period of stagnation or even decline lasted is difficult to determine. Certainly, by the beginning of the Tudor period (1485) the steady growth of population and the economy must once again have caused the road network to grow. Medieval roads were clearly sufficient for the amount of traffic on them, and this is emphasised by the fact that virtually no new roads were built. It is not until some years after the dissolution of the monasteries (1536–40), which had affected much of what little road maintenance was done, and the more rapid growth of the economy in the mid-sixteenth century, that the system started to break down; travellers started to complain bitterly about the state of the roads, and in 1555 Parliament had to introduce legislation for their upkeep.[53] It was thus not until the middle of the sixteenth century that the fine balance between the amount of traffic and the medieval roads' ability to maintain themselves was broken, and the system which had come into use over the whole medieval period in response to the demand from all kinds of travellers finally became unable to cope with the sheer volume of traffic.

Notes

1. R.A. Dodgshon and R.A. Butlin, *An Historical Geography of England and Wales* (Academic Press, London, 1978), pp. 103, 145; C. Platt, *Medieval England* (Routledge and Kegan Paul, London 1978); M.M. Postan, *The Medieval Economy and Society* (Weidenfeld and Nicolson, London, 1972); J.C. Russell, *Medieval Regions and their Cities* (David and Charles, Newton Abbot, 1972).

2. W. Cunningham, *The Growth of English Industry and Commerce during the Early and Middle Ages* (Cambridge University Press, Cambridge, 1915); R.A. Donkin and R.E. Glasscock in H.C. Darby (ed.), *A New Historical Geography of England before 1600* (Cambridge University Press, Cambridge, 1976), pp. 116-9, 174-5; J.E.T. Rogers, *Agriculture and Prices in England, 1259-1400*, London, 1866; J.A. Raftis, *Tenure and Mobility: Studies in the Social History of the Medieval English Village* (Pontifical Institute of Medieval Studies, Toronto, 1964).

3. J.J. Jusserand, *La Vie Nomade et les Routes d'Angleterre du XIVe Siècle* (Paris, 1884), pp. 32-7; M.M. Postan, *Medieval Trade and Finance* (Cambridge University Press, Cambridge, 1973), pp. 116-19; L.F. Salzman, *English Trade in The Middle Ages* (Oxford University Press, Oxford, 1931), pp. 185-90.

4. C.T. Flower, 'Public Works in Medieval Law', *Selden Society*, vol. 32 and vol. 40 (1915, 1923), vol. 2, p. xvi.

5. I.H. Adams, *Agrarian Landscape Terms* (Institute of British Geography, Special Publication 9, 1976), pp. 129-32; S. and B. Webb, *The Story of The King's Highway* (Longmans, London, 1913), pp. 5-6.

6. B.P. Hindle, 'Seasonal Variations in Travel in Medieval England', *Journal of Transport History*, vol. 4 (1978), pp. 170-8; C.A.J. Armstrong, 'Some Examples of the Distribution and Speed of News in England at the time of the Wars of the Roses' in R.W. Hunt, W.A. Pantin and R.W. Southern (eds.), *Studies in Medieval History presented to F.M. Powicke* (Oxford University Press, Oxford, 1948), pp. 429-54.

7. J.F. Willard, 'Inland Transportation in England during the Fourteenth Century', *Speculum*, vol. 1 (1926), pp. 361-74.

8. F.M. Stenton, 'The Road System of Medieval England', *Economic History Review*, vol. 7 (1936), pp. 1-21; E.J.S. Parsons, *The Map of Great Britain, c. AD 1360, known as the Gough Map* (The Bodleian Library, Oxford and The Royal Geographical Society, London, 1958).

9. F.M. Stenton, 'The Road System of Medieval England', p. 6.

10. Ibid., p. 21.

11. B.P. Hindle, 'The Road Network of Medieval England and Wales', *Journal of Historical Geography*, vol. 2 (1976), pp. 207-21.

12. M.W. Beresford and J.K.S. St Joseph, *Medieval England: An Aerial Survey* (Cambridge University Press, Cambridge, 1979), esp. pp. 273-84.

13. C. Taylor, *Fieldwork in Medieval Archaeology* (Batsford, London, 1974), pp. 82-3, 90, 93, 142.

14. C. Platt, *The English Medieval Town* (Secker and Warburg, London, 1976), pp. 48-50.

15. Flower, 'Public Works in Medieval Law'.

16. Ibid., vol. 2, pp. 207-8.

17. H. Poole and J.P. Gilson, *Four Maps of Great Britain designed by Matthew Paris about AD 1250* (British Museum, London, 1928).

18. J.B. Mitchell, 'The Matthew Paris Maps', *Geographical Journal*, vol. 81 (1933), pp. 27-34.

19. B.P. Hindle, 'The Towns and Roads of the Gough Map (c. 1360)', *The Manchester Geographer*, vol. 1 (1980), pp. 35-49.

20. Parsons, *The Map of Great Britain, c. AD 1360*.

21. B.P. Hindle, 'The Towns and Roads of the Gough Map'; Parsons, *The Map of Great Britain, c. AD 1360.*

22. Giraldus de Barri, *The Itinerary of Archbishop Baldwin Through Wales,* edited by R.C. Hoare, (1806); W. Rees, *An Historical Atlas of Wales* (Faber and Faber, Cardiff, 1951), plate 37.

23. B.P. Hindle, 'A Geographical Synthesis of the Road Network of England and Wales', unpublished PhD thesis, University of Salford, 1973, pp. 94–100.

24. G.H. Martin, 'Road Travel in the Middle Ages', *Journal of Transport History,* vol. 3 (1976), pp. 159–78.

25. Exchequer Accounts, Miscellaneous, 309/29.

26. J.E.A. Jolliffe, 'The Chamber and Castle Treasures under King John' in R.W. Hunt, W.A. Pantin and R.W. Southern (eds.), *Studies in Medieval History presented to F.M. Powicke,* pp. 118–19.

27. T.D. Hardy, 'Itinerarium Johannis Regis Angliae', *Archaeologia,* vol. 22 (1829), p. 143.

28. H. Gough, *Itinerary of King Edward the First* (Paisley, 1900), vol. II, p. 64.

29. Hindle, 'The Road Network of Medieval England and Wales', pp. 213–16; 'A Geographical Synthesis of the Road Network of England and Wales', pp. 59–92.

30. S. Uhler, 'English Customs Ports, 1275–1343', unpublished BPhil dissertation, University of St Andrews, 1977, pp. 218–78.

31. Ibid., p. 259.

32. R.A. Donkin, *The Cistercians: Studies in the Geography of Medieval England and Wales* (Pontifical Institute of Medieval Studies, Toronto, 1978), pp. 138–43.

33. P.D.A. Harvey, *A Medieval Oxfordshire Village* (Oxford University Press, Oxford, 1965), pp. 98–104.

34. J.C. Russell, *British Medieval Population* (University of New Mexico Press, Alberqueque, 1948), pp. 140–3.

35. Hindle, 'A Geographical Synthesis of the Road Networks of England and Wales', pp. 113–39.

36. Hindle, 'The Road Network of Medieval England and Wales', pp. 217–19.

37. Beresford and St. Joseph, *Medieval England: An Aerial Survey,* pp. 273–84.

38. Ibid., p. 273.

39. Ibid., pp. 276–7; C. Taylor, *Fieldwork in Medieval Archaeology,* p. 85.

40. Beresford and St Joseph, *Medieval England: An Aerial Survey,* pp. 278–81.

41. C. Taylor, *Roads and Tracks of Britain* (Dent, London, 1979), pp. 115–19; see also O.G.S. Crawford, *Archaeology in the Field* (Ordnance Survey, London, 1953), pp. 67–86.

42. C. Taylor, *Roads and Tracks of Britain,* pp. 120–4.

43. Beresford and St Joseph, *Medieval England: An Aerial Survey,* pp. 282–4; H.C. Darby, *The Medieval Fenland* (Cambridge University Press, Cambridge, 1940), pp. 106–18.

44. B.P. Hindle, 'Medieval Roads in the Diocese of Carlisle', *Trans. Cumberland and Westmorland Antiquarian and Archaeological Society,* vol. 77 (1977), pp. 83–95. See also B.P. Hindle, *Lakeland Roads* (Dalesman, Clapham, 1977).

45. B.P. Hindle, 'A Geographical Synthesis of the Road Network of England and Wales', pp. 161–72.

46. W. Harrison, 'Pre-Turnpike Highways in Lancashire and Cheshire', *Lancashire and Cheshire Antiquarian Society,* vol. 9 (1891), pp. 101–34; H.J. Hewitt, *Medieval Cheshire* (Manchester University Press, Manchester, 1929), ch. 4.

47. D. Sylvester, 'Cheshire in the Dark Ages', *Historic Society of Lancashire and Cheshire,* vol. 114 (1962), pp. 1–22.

48. W.B. Crump, 'Saltways from the Cheshire Wiches', *Lancashire and Chesire Antiquarian Society*, vol. 54 (1939), pp. 84–142.

49. Harrison, 'Pre-Turnpike Highways in Lancashire and Chesire'; Hewitt, *Medieval Cheshire*.

50. Gough, *Itinerary of King Edward the First*; T. Craib, 'Itinerary of Henry III, 1215–1272', (typescript) Public Record Office, Round Room Press, 22/44 (1923).

51. J.C. Russell, *British Medieval Population*.

52. W. Harrison, 'Description of Britain', *Holinshead's Chronicles* (1586), vol. I, pp. 414–21; W. Smith, *The Particular Description of England* (1588); J. Ogilby, *Britannia* (1675), vol. i, pp. 23, 37, 57, 63, 90; R. Morden, *The County Maps of William Camden's Britannia, 1695* (David and Charles, Newton Abbot, 1972).

53. Harrison, 'Description of Britain', vol. i, p. 192; C.W. Scott-Giles, *The Road Goes On* (Epworth Press, London, 1946), pp. 63–103.

8 CONCLUSION

Leonard Cantor

The contributors to this book have undertaken a common task, namely the reconstruction of specific aspects of a past landscape and an analysis of the changes that took place in that landscape through a considerable period of time. Such studies are valuable for several reasons: they have an intrinsic interest; they help to deepen our understanding of our landed heritage; they illuminate the ways in which cultural activities flourish and decay; they provide a basis of knowledge and stimulate the development of techniques which can be applied to the fashioning of geographies of the future; and they are essential to a sympathetic understanding of today's landscape and the need to identify and conserve individual features and possibly entire landscapes of historical interest and aesthetic value.[1]

As we have seen, the landscape of medieval England was in many respects more varied and more intensively used than that of the present day. Certainly by the thirteenth century, it was a 'busy' landscape, like a Brueghel painting peopled with peasants at work in the fields and richly endowed with flora and fauna. For example, Kershaw writes of the district of Craven in the north-west of the West Riding of Yorkshire:

> The traveller today in Craven sees scarcely an acre of arable under cultivation. Sheep and dairy herds have taken over completely. In the Middle Ages the landscape must have looked quite different, with extensive areas given over to crop growing, mainly for subsistence, and with an army of labourers reaping the crops in fields now occupied only by a few grazing animals.[2]

In addition to the intensive exploitation of areas of arable farming, and also pastoral farming, tracts of land such as heather, fens, moorlands and marsh were systematically used,[3] and the settlement pattern of village and town, linked by roads and tracks, was already well established. In our task of reconstructing this rich variety of land use, we have drawn extensively on two essential and prime sources of evidence: documentary archival material, including maps and air photographs; and the English landscape itself.

218

However, despite the wide range of documentary material now available to us, it still constitutes only a partial record of medieval England. This is inevitable as we are dealing with a period of time, between 500 and 900 years ago, which was largely pre-literate and in which relatively few records were kept. Moreover, as even fewer have survived, there is always a danger of drawing general conclusions from documents which are unusual because of the very fact of their survival and may or may not portray a picture widely representative of the times with which they deal. Certainly, one of the perennial fascinations of documentary evidence is that, in Prince's words, it 'may be read and reread in an infinite variety of ways'.[4] As for the English landscape itself, it constitutes a palimpsest on which each generation inscribes its own impressions and removes some of the marks left by previous generations.[5] Man's activities during the Middle Ages have made a major contribution to that palimpsest and have left us numerous relict features, albeit on a decreasing scale as modern farming methods and industrial developments obliterate all traces of what has gone before. Nevertheless, as the authors of this book have clearly demonstrated, much remains to be identified and recorded. Fortunately, there has been an upsurge of interest in landscape history in the last twenty-five years, during which the work of professional archaeologists and academic historical geographers has been supplemented by that of enthusiastic groups of people such as the Medieval Village Research Group and the Moated Sites Research Group, operating mainly in their own localities to identify and record specific landscape features.

For all of us, however, amateurs and professionals alike, we have as yet only, literally, skimmed the surface of medieval England. Whether it is the mapping of moated sites or medieval hunting parks, or tracing the remains of ridge and furrow, or recording our inheritance of medieval buildings, far more remains to be done than has yet been achieved. The English landscape has been described as the finest compact between man and the land ever made. Alas, this compact has been severely damaged in recent years by modern technology so that the importance of identifying and recording, and hopefully preserving, our medieval landscape heritage has never been greater. This endeavour is not for the professional historical geographer alone, but offers rewards and pleasures to all who are prepared to make the effort. In the words of the late Lucian Febvre, the historian, and by extension the historical geographer, 'is not he who knows, but he who seeks'.[6]

NOTES ON CONTRIBUTORS

Peter Bigmore Senior Lecturer in Geography and Planning, Middlesex Polytechnic

Leonard Cantor Schofield Professor of Education and Head of Department of Education, Loughborough University of Technology

Brian Paul Hindle Lecturer, Department of Geography, University of Salford

Trevor Rowley Staff Tutor in Archaeology, Department for External Studies, University of Oxford

Michael Williams Fellow and Tutor of Oriel College and University Lecturer, School of Geography, University of Oxford

INDEX

abbeys 144, 146; in forests 59; in Somerset 103; *see also* Crowland, Fountains, Glastonbury, Meaux, monasteries

Aberg, F.A. 138, 141

aerial photographs 171, 207, 218

Air Photographic Unit, Cambridge University 172

Anglo-Saxons: hunting 57; place-names 164

Appleton-le-Moors (North Yorks) 167-8

Arden, Forest of 89, 109-10

Ashdown Forest, warrens in 83

assarting 39, 42, 92-4, 108; in forests 65, 112

Augustinians 147

Benedictines 144, 147, 150

Beresford, M.W. 87, 155, 170, 174, 175, 188n1, n4, 190n42, 195, 206, 207

Bicester, Oxon 179, 180

Black Death, 1348-9 19, 77, 89, 119, 120, 150, 169, 214; *see also* plagues

Boarstall, Bucks, 1444 map of 28

bridges 195

Bristol 155, 184, 187

Brown, R. Allen 126, 137

Bury St Edmunds 22, 176, 177, 183, 186

Cannock, Forest of 66, 77

Cannock Chase 72-3

Canterbury 187, 195, 196, 199

Carburton (Notts), infield–outfield system 37-8

Castles 19, 126-37; decline of 132-3; distribution of 128-9, 135-7; functions of 136-7; in Leicester-shire 136; keep and bailey 130; motte-and-bailey 127-9, 131; ringworks 127, 152n27; tower houses 131; *see also* Tutbury Castle, Staffs

causeways 195, 209; *see also* roads

Celtic fields 33, 34

chases 56, 70-3; *see also* Chases of Cannock, Cranborne, Leicester

Cheshire, forests of 63-6

Chester 212, 213

Cholwich Farm, Dartmoor, Devon 114-15

Cistercians 106, 114, 116, 118, 143-4, 146, 147, 150, 151, 202

climatic changes 20, 119, 169

cloth manufacturing *see* textile industry

Cluniac houses 146

coal mining 23, 48

common land 30; *see also* waste

coneries *see* warrens

Corpus Christi estate of Whitehall, Tackley (Oxon) 32, 33

Cranborne Chase 73

Crowland Abbey (Lincs) 96, 110-11, 144

Darby, H.C. 17, 91, 121, 157, 195

deer: in chases 70; in forests 57, 63, 65; in parks 75-6

deer-folds 76

deserted medieval villages 48, 170; *see also* villages, abandonment of

Domesday Book 17, 79, 91, 95, 129, 158, 160, 164, 176

Domesday England 17-18, 157-8

domus defensabiles (defensible houses) 133

Dorset, common fields in 21

Dover 196, 199

doves 150

dykes, in Fenlands 99-100, 101-2

early Middle Ages 18

enclosed fields 38-41

enclosure of fields 21, 31, 164; *see also* enclosed fields

Exe–Tees line 21-2

Exeter 22, 155, 186

Fenlands 19, 40, 87, 89, 97-102, 107, 119, 144, 195, 209; *see also* dykes, villages

fens *see* marshlands

fisheries, in marshlands 95

fish-ponds 78, 138, 146

222

Milton Keynes UK
Ingram Content Group UK Ltd.
UKHW040101071024
449327UK00019B/709

9 780367 747541